ROWAN UNIVERSITY
CAMPBELL LIBRARY
201 MULLICA HILL RD.
GLASSBORO, NJ 08028-1701

High Performance ASIC Design: Using Synthesizable Domino Logic in an ASIC Flow

Presenting methodologies for high speed ASIC design developed over several years in industry, this practical book covers issues related to the use of domino logic in an automated framework, and brings together all the knowledge needed to apply them in practice.

An overview of design techniques used to achieve high speed in ASIC designs is followed by chapters describing the design and characterization of domino logic standard cell libraries and an advanced domino logic synthesis flow. Actual results achieved by using automated domino logic design techniques, including silicon measurements, are presented to validate the methodology, whilst real-world design examples, such as the implementation of the execution unit of a microprocessor and Viterbi decoder, show how the techniques are applied in practice. This book is ideal for graduate students and researchers in electrical and computer engineering, and also for circuit designers and EDA engineers in industry.

RAZAK HOSSAIN is a Senior Principal Engineer at STMicroelectonics Inc., San Diego, California, where he has worked since 2000 on high-speed ASIC chips and design methodologies. He earned his Ph.D. in Electrical Engineering from the University of Rochester, New York, in 1995, after which he worked on structured custom circuit design at Mentor Graphics Corporation, Warren, New Jersey.

High Performance ASIC Design

Using Synthesizable Domino Logic
in an ASIC Flow

Razak Hossain
STMicroelectronics

CAMBRIDGE UNIVERSITY PRESS
Cambridge, New York, Melbourne, Madrid, Cape Town, Singapore, São Paulo, Delhi

Cambridge University Press
The Edinburgh Building, Cambridge CB2 8RU, UK

Published in the United States of America by Cambridge University Press, New York

www.cambridge.org
Information on this title: www.cambridge.org/9780521873345

© Cambridge University Press 2008

This publication is in copyright. Subject to statutory exception
and to the provisions of relevant collective licensing agreements,
no reproduction of any part may take place without
the written permission of Cambridge University Press.

First published 2008

Printed in the United Kingdom at the University Press, Cambridge

A catalog record for this publication is available from the British Library

ISBN 978-0-521-87334-5 hardback

Cambridge University Press has no responsibility for the persistence or
accuracy of URLs for external or third-party internet websites referred to
in this publication, and does not guarantee that any content on such
websites is, or will remain, accurate or appropriate.

Contents

	Preface		*page* vii
	Abbreviations		ix
1	**An introduction to domino logic**		**1**
	1.1	CMOS and NMOS	1
	1.2	Domino logic circuits	5
	1.3	Clocking domino logic	12
	1.4	Summary	15
2	**High-speed digital design**		**18**
	2.1	Microprocessors since 1989	18
	2.2	Microarchitectures for high speed	22
	2.3	Designing and using high-speed memories	31
	2.4	What to remember if applying domino logic	35
3	**Domino logic library design**		**37**
	3.1	High-speed digital circuit design	37
	3.2	An introduction to standard cells	42
	3.3	Designing a high-performance standard cell library	45
	3.4	Circuit design of domino logic cells: a qualitative approach	48
	3.5	Circuit design of domino logic cells: a quantitative approach	51
	3.6	Characterizing domino logic-compatible registers	63
	3.7	Layout of domino logic standard cells	65
	3.8	Timing models for domino logic cells	66
4	**Domino logic synthesis**		**70**
	4.1	Introduction to domino logic synthesis	70
	4.2	Unate transform	73
	4.3	Phase assignment	75
	4.4	Phase-assignment rules	77
	4.5	An example domino synthesis flow	86
	4.6	Schematic capture of domino designs	106

5	**Circuits designed with domino logic in an ASIC flow**		**108**
	5.1	Introduction	108
	5.2	Domino integer execution unit	108
	5.3	A synthesized domino logic DSP core	119
	5.4	A synthesizable domino logic Viterbi add–compare–select (ACS) test chip	121
	5.5	Intel's published domino logic synthesis flow	124
	5.6	Conclusions	126
6	**Evolution of domino logic synthesis**		**127**
	6.1	The state of digital ASIC design methodologies	127
	6.2	Process trends and domino logic	128
	6.3	Clocking methodology for domino circuits	130
	6.4	Synthesizing other dynamic logic families	132
	6.5	Flow improvements for domino synthesis	137
	6.6	The case for domino logic synthesis	141
	Index		143

Preface

This book stems from my experience over the last few years in designing high-speed digital logic using ASIC design flows. I discovered that while it is possible to significantly improve performance in ASIC implementations with deep pipelining and careful physical design, a speed penalty still had to be paid due to their exclusive use of static logic. This spurred an interest in using domino logic with automated synthesis and place and route tools. This book documents my experiences in automating the use of domino logic, and shows that despite the challenges entailed in the process, it is possible to use domino logic with industry-standard ASIC tools and achieve a significant speed improvement in the process.

Engineering is a group activity. The development of our domino logic synthesis system was possible due to the collaboration of many intelligent, enthusiastic, and dedicated co-workers whose contributions I must acknowledge. First of all I would like to thank my two chapter co-authors, Tommy Zounes and Bernard Bourgin. In addition to being gifted and hard-working engineers, Tommy and Bernard have also always been very generous with their knowledge and time, allowing all of their co-workers, including me, to learn a great deal from them. The domino logic library was possible due to the talents and efforts of Scott Anderson, Shaun Forsting, Judy Alvarez-Gallardo, Roger Boates, Michael Lin, and Juneho Park, who helped design the schematics and also contributed to the myriad other tasks involved with taping out a number of chips. Scott, armed with a contagious optimism, also helped me document our early experiences with using domino logic. Shaun Forsting converted the schematics into very efficient layouts across a number of different CMOS processes. During the early years of the domino logic project we were joined by two engineers from Italy: Fabrizio Viglione and Marco Cavalli. They both worked on the first domino chips with great enthusiasm and effectiveness. Fabrizio subsequently took the first stab at implementing our approach to synthesizing domino logic. From France we were later joined by the affable Leonardo Valencia, who with Cyril Adobati and Robin Wilson completed the first design that used a fully synthesizable domino logic flow. Roy Mader and Boris Andreev worked on the project as summer interns. Roy subsequently became a much-valued permanent member of our group and led us in overcoming many of the onerous challenges involved in pushing domino designs through automated place and route flows.

People work effectively only in a supporting environment. I would like to thank our manager Naresh Soni for encouraging and supporting us in our work in domino synthesis, as well as Joel Monnier, who led STMicroelectronics' Central Research and Development organization. They provided us with the extraordinary luxury of being allowed to innovate in an autonomous manner. Nick Richardson, who later led the group, continued in this fine tradition and also provided more specific technical advice on matters related to logic and architecture. In addition, I must thank the many others in STMicroelectronics who supported us in our work on domino logic, including: Philippe Magarshack, Jean-Pierre Schoellkopf, Sylvain Kritter, Heloise Tupin, Damien Croain, Alain Chion, Samala Sreekiran, Sanjay Bulusu, Ezio Iacazio, and Marco Gregori.

I would like to thank my wonderful parents, Mosharaff and Inari Hossain, who have encouraged me throughout this endeavor, and more broadly, instilled in me a love of books and learning. Finally I must thank my beloved wife, Zakia Chowdhury, whose support allowed me to write this book. I dedicate this book to her and my two delightful sons, Farhan and Ishraq.

Razak Hossain
San Diego, CA

Abbreviations

ASIC	application-specific integrated circuit
CMOS	complementary metal oxide semiconductor
CSA	carry save adder
CTO	clock tree optimization
DRC	design rule check
DSPF	detailed standard parasitic format
DVD	digital video disc
ECO	engineering change order
EDA	electronic design automation
FET	field effect transistor
fF	femtofarad
FO4	fan-out of four
GHz	gigahertz
HDL	hardware description language
LFSR	linear feedback shift register
LSB	least significant bit
LVS	layout versus schematic
MHz	megahertz
MIPS	million instructions per second
MPC	minimum physical constraints
MPWH	minimum pulse width high
MPWHO	minimum pulse width high overlap
MPWL	minimum pulse width low
MSB	most significant bit
MUX	multiplexer
NMOS	n-channel metal oxide semiconductor
PLL	phase locked loop
PMOS	p-channel metal oxide semiconductor
PUT	pin under test
PVT	process, voltage, and temperature
QoR	quality of results
RC	resistor capacitor circuit
RF	radio frequency

RISC	reduced instruction set computer
RTL	register transfer level
SDC	Synopsys design constraints
SOC	system-on-chip
SRAM	static random access memory
TAT	turnaround time
VCO	voltage-controlled oscillator
VLSI	very large scale integration
XNOR	exclusive NOR
XOR	exclusive OR
μm	micrometer

1 An introduction to domino logic

1.1 CMOS and NMOS

By the late 1970s complementary metal oxide semiconductor (CMOS) started to become the process of choice for digital semiconductor designs. CMOS had originally been proposed by Frank Wanlass in 1963 as a low standby power technology, since CMOS logic gates dissipate almost no power when the inputs to the gate do not change [1]. This follows as CMOS contains both PMOS field effect transistors (FETs), which can efficiently drive a high voltage, or logic one value, and NMOS transistors, which are good at driving a zero voltage. The presence of complementary transistors allows CMOS logic gates to be implemented so that the output voltage level is connected to the power or ground line, but not both. This ability to avoid contention ensures that if the inputs are not changing, then no power is dissipated. This was a major advantage of CMOS over the other manufacturing processes then available, which dissipated constant leakage or bias currents.

In Figure 1.1 the schematic representation of a CMOS static NAND logic gate is shown. The logic gate has two inputs A and B. A high logic value at inputs A and B turns on transistors MN1 and MN2, while turning off transistors MP1 and MP2. This causes the output Z to be low. When either input A or B is off, however, the path to the ground line is ruptured, with a path to the power supply (by convention called Vdd) being established. This causes Z to rise. While a NAND gate represents a simple function, it does show how contention between the power and ground supplies can be avoided in CMOS circuits. This lack of contention means that when the inputs to a CMOS circuit do not change, often called a standby or idle state, almost no power dissipation occurs, except for a small leakage current which flows through the transistors due to the imperfect manner in which a MOSFET acts as a switch (due to the relentless scaling in the physical dimensions of CMOS processes, driven by the cost advantages of having a smaller silicon area for digital functions, MOS transistors have become less perfect switches, leading to greater leakage current).

The fact that CMOS logic would lead to substantial power savings was apparent to its inventor Frank Wanlass, who in 1963 was working at Fairchild Semiconductor. Wanlass attempted to prove the viability and technical advantages of CMOS with a monolithic implementation of the technology [2]. When this proved infeasible, he proved the concept with discrete transistors. His CMOS implementations reduced standby power by six orders of magnitude over equivalent bipolar and PMOS implementations [2]. While

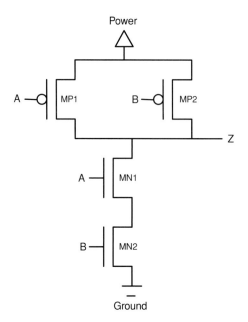

Figure 1.1. A static CMOS two-input NAND cell.

impressive, this advantage of CMOS would not prove decisive for many years. Early monolithic designs were very small, with the standby power consequently being very small as an absolute quantity. The inferior maturity of MOS transistors meant that in the 1960s, bipolar logic raced ahead of MOS transistors in applications. Transistor–transistor logic (TTL) and emitter-coupled logic (ECL), developed in 1962 and 1966, respectively, provided effective digital design techniques for bipolar transistors in the rapidly increasing semiconductor industry, which by 1962 had surpassed a billion dollars in annual sales [2]. The major user of CMOS in its early years was the watch industry, where battery life was a more important attribute than speed [3]. Starting in the 1970s, MOS technology began to mature rapidly, with much of the early industrial development being driven by Intel, then a small Silicon Valley company. In 1971 Intel released the 4004, the world's first microprocessor. The 4004 was built using a 10 µm line width PMOS transistor and used 2300 transistors running at 108 kHz [4]. In 1974 Intel released the 8-bit 8080, manufactured in a 6 µm NMOS process. The chip ran at 2 MHz and had 6000 transistors. Yield and cost concerns at the time ensured manufacturers preferred to use a single type of MOS transistor. Since NMOS transistors were faster than PMOS ones, due to the higher mobility of electrons over holes, the move to an NMOS process was natural.

Figure 1.2 shows the schematic implementation of a NAND gate using NMOS transistors only. The PMOS transistors MP1 and MP2 shown for the CMOS implementation in Figure 1.1 are removed here and replaced by a resistor, R1. This conceptual resistor is actually implemented by a depletion mode NMOS transistor [5]. The NMOS NAND gate output is at Vdd, or a logic one value, when either of the inputs, A or B, is low. When input A and input B are both high, the output is driven low. The current-driving

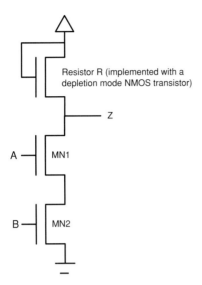

Figure 1.2. An NMOS two-input NAND cell.

ability of pull-down NMOS transistors must be much greater than that of the pull-up resistor. This ensures that the output can be driven to a low voltage at the cost of higher power dissipation. In addition to the standby power dissipation, NMOS circuits tend to be slower than equivalent CMOS circuits. This is due to the need for a weak pull-up resistor, which results in very slow low-to-high transitions. While these disadvantages may make NMOS appear to be unappealing, NMOS designs are more compact than CMOS circuits. Figure 1.2 uses only two transistors and a resistor, compared with the four transistors needed by a CMOS design. Since the pull-up resistor is implemented by another NMOS MOSFET, the NMOS design uses fewer transistors and a simpler process than the CMOS design. The need to move to CMOS therefore arose only when the integration level on integrated circuits (ICs) made the large standby power on the NMOS design unacceptable. For Intel this transition occurred in 1978, when the 8088/8086 family of microprocessors was introduced (the designs were almost identical to the 8088, having an 8-bit bus while the 8086 has a 16-bit bus). With 29,000 transistors and a clock rate of 5 to 10 MHz, the 8086 dissipated 1.5 W. This exceeded the 1 W per chip power limit for plastic packaging. Increases in integration levels meant that a 32-bit processor would dissipate 5 to 6 W, leading to severe reliability problems [6]. The CMOS version of the 8086, the 80C86, consumed only 250 mW [6]. The ability of CMOS to reduce power dissipation with increasing integration meant that it rapidly emerged as the technology that could best utilize fabrication advances. It is an advantage that CMOS maintains till today (2007), with the overwhelming majority of digital IC designs in the world being manufactured in CMOS, and the increased convergence of systems onto chips leading CMOS to make strong inroads into analog and radio frequency (RF) designs. In 1980, Intel's 8088 was chosen by IBM as the microprocessor for its personal computer (PC) [4], a step that would lead to Intel becoming, within a few years, the largest semiconductor company in the world, with its semiconductor revenues far exceeding that of IBM itself. The rest, as they say, is history.

As semiconductor manufacturing progressed, the largest challenge to the nascent industry was the ability to design and verify designs using the increasing number of transistors available. This need was met by the development of a new field of software, often closely tied to dedicated hardware in its early years, called electronic design automation (EDA). EDA developments started in the 1960s, with software developed in-house by different semiconductor companies. In the early years of EDA the most common tools developed were circuit and logic simulators, which allowed designers to verify the expected functionality of a design before manufacturing. Alberto Sangiovanni-Vincentelli states in his excellent history of EDA (*The Tides of EDA*) that the early tools had limited loyalty due to the perceived limited value-added of the tools [7]. By the late 1980s continued developments in EDA had resulted in the development of logic synthesis, which could map a register transfer level (RTL) description to a set of standard cell gates and memory instances, and automated physical design tools, which could physically instantiate and route the wires needed to complete the physical design. These tools led to a marked improvement in productivity, allowing digital designs to be quickly implemented based on a higher abstraction level, behavioral RTL description [7]. The increasing complexity of EDA tools, along with the realization of their tremendous usefulness, led to the rise of independent EDA companies and a rapid reduction in EDA tool development within semiconductor companies. By the end of 2006, the EDA industry had a total available market (TAM) of 5.3 billion dollars [8], which is about 2% of the worldwide semiconductor TAM of 260 billion dollars [9]. The success of CMOS manufacturing technology, along with the availability of powerful EDA tools, allowed for the widespread penetration of electronics into a multitude of applications.

People who are drawn into, and ultimately stay in, engineering are generally somewhat private people. Our work must, by definition, be cooperative, but the bread and butter of our daily tasks tends to be very solitary exercises. I am aware that such an audience feels extremely uncomfortable with broad, historical utterances, reminding them of overly optimistic forecasts they have had to sit through in darkened conference rooms with a roll of the eye and a quick, knowing smile to a colleague. Still, I feel compelled to state the following: I have no doubt that looking back from the future, the most important historical event of our age will be the development and promulgation of digital technology. It will also be seen as a profoundly positive development. I believe that all of us working in this field should be proud of our achievements. I have said my piece and now return to the theme of digital ASIC design.

It may have been assumed that the emergence of ASIC design methodologies would displace all other techniques for implementing digital CMOS logic. This has not happened, as many digital designs have specific needs that cannot be achieved by using standard ASIC techniques. In recent years the capabilities of ASIC tools have increased greatly, largely due to the tremendous competition among the companies in the field. Many logical and behavioral optimizations that previously had to be hand-coded for efficient implementation are now automatically incorporated in the synthesis tools. The two most common benefits of custom design are its ability to optimize across the different

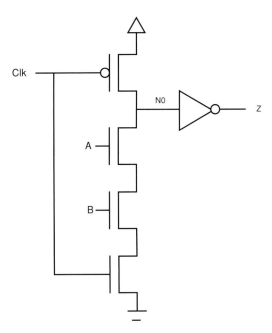

Figure 1.3. A CMOS domino logic two-input AND gate.

levels of abstractions in the ASIC design framework and the opportunity it provides for using logic families other than standard static logic. The first of these advantages relates to the sequential approach that an ASIC design methodology uses, by which standard cell library development, logic synthesis, and physical design are broadly separate processes. It is true that synthesis tools are increasingly aware of physical design constraints and physical design tools can perform logic optimizations. This is still very far away from them providing a common design framework. For example, if a very high-speed design can be made fast and small using a very specific cell that is instantiated and placed in a particular manner, it is improbable that an automated design flow will be able to reach that exact solution. If the library does not have the particular standard cell in its library, then the automated solution obviously cannot use it.

The second advantage of custom design is that it can utilize certain logic families, specifically dynamic logic, that automated design frameworks have not traditionally been able to support. In this book we describe our experiences in incorporating domino logic into an ASIC design flow. This journey starts with a short description of what domino logic is.

1.2 Domino logic circuits

A picture, it has been said, is worth a thousand words. We therefore begin our description of domino logic with Figure 1.3, which shows the schematic representation of a domino logic two-input AND gate.

The AND gate shown in Figure 1.3 can be used to illustrate the functionality, the speed advantage, and also some of the challenges involved in using this logic family. In Figure 1.3 it can be seen that the two functional inputs, A and B, are also attended by the clock signal, Clk. At first glance this may seem strange, since an AND gate should be a purely combinational circuit, which unlike latches and flip-flops does not require the presence of the clock signal. Domino logic is, however, a clocked logic family, which means that every single logic gate has a clock signal present. When the clock signal turns low, node N0 (which is called the evaluation or internal node – some authors refer to it as the dynamic node) goes high, causing the output of the gate to go low. This represents the only mechanism for the gate output to go low once it has been driven high. The operating period of the cell when its input clock and output are low is called the precharge phase or cycle. The next phase, when the clock is high, is called the evaluate phase or cycle. During the evaluate phase the output of the domino AND cell can go high provided that both inputs A and B are high, which causes the evaluation node, N0, to be driven to a low value. The evaluate phase is the functional operating phase in domino cells, with the precharge phase enabling the next evaluate phase to occur. The appropriate application of the clock signal ensures that the critical path in domino cells only traverses through cells in the evaluate phase. One of the advantages of domino logic over static logic can also be garnered from the schematic in Figure 1.3. Since the domino cell only switches from a low to a high direction, there is no need for the inputs A and B to drive any pull-up PMOS transistors. The lack of a PMOS transistor means that the effective transistor width that loads down a previous stage of logic, for a particular current drive, favors domino over static logic. This is critical since the key to high speed is ensuring that a speed advantage can be gained without loading down the cell greatly [10]. For example, if a design is constructed with a set of cells with transistors of a certain size, replacing the transistors in every cell with ones ten times larger will almost certainly lead to a design that is faster. Provided that the initial design is properly sized, i.e., without weak cells which have very long rise or fall times, the new design will not, however, be ten times faster. The reason for this is that, while the drive strengths of each cell have increased by a factor of 10, the output loading due to the input transistor capacitance seen by each cell has also increased by approximately a factor of 10. Since larger cells are now used in the design, its area will be larger, leading to greater wiring capacitance. Thus, while speed gains can be achieved by optimizing cell drives, the indiscriminate increase in drive strengths tends to limit the improvement in speed due to the increased self-loading.

In order to see how domino logic alters the relationship between input capacitance and output drive strength, compared with an equivalent static cell, the reader is directed to Figure 1.4. A static buffer is shown in the figure, with input PMOS and NMOS transistor widths of 2 µm and 1 µm, respectively. Assuming that the gate capacitance of a PMOS and NMOS transistor is the same per unit micrometer of transistor width, the total load seen by the cell driving the buffer is 3 µm of transistor gate width. For a domino cell it is possible to construct a buffer with the same drive strength but which has only 1 µm of transistor width as input capacitance. Alternately, with the same input capacitance it is possible to build a stronger and faster domino buffer. This is shown in Figure 1.4,

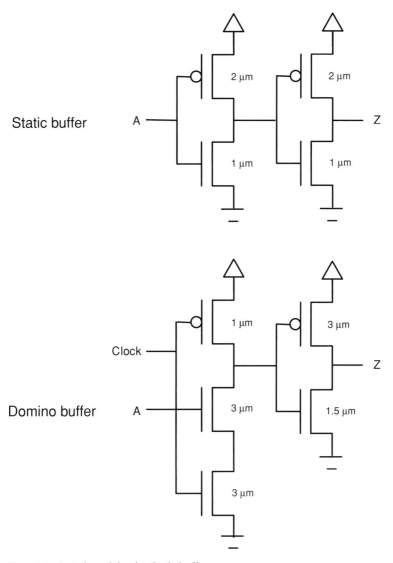

Figure 1.4. A static and domino logic buffer.

where a domino buffer with input transistor width of 3 μm is shown. This particular domino buffer is called a footer transistor, i.e., a series NMOS transistor connected to the clock. It is possible to use domino cells with and without footer transistors, although the absence of the footer transistor makes the design more complicated. Since the footed clock transistor adds to the resistance of the pull-down path, the two 3 μm transistor widths are considered roughly equivalent in terms of drive strength to a single 1.5 μm transistor. This follows, as the resistance of a MOS transistor is inversely proportional to its gate width, and as the total resistance of a set of resistors in series is equal to the sum of the resistors. Since, for the static buffer, 1 μm of NMOS transistor length is driving a 2 μm PMOS and a 1 μm NMOS transistor, the effective 1.5 μm of NMOS for the domino cell has 50% greater drive strength. Thus, for the same input capacitance a domino cell can result in greater output drive strength than a static equivalent.

There are a few points to note. Firstly, the degradation in drive strength due to the addition of the footer transistor in a domino buffer is worse than in other domino cells with more than two NMOS transistors in series. For example, in a three-input domino AND cell, the number of NMOS transistors in series goes from three to four when the footed transistor is considered. This is much better than a domino buffer where a doubling in the height of the NMOS series stack occurs. Secondly, the PMOS pull-up transistor for the domino cell in Figure 1.4 is shown as 1 µm. The actual size of the PMOS transistor will depend on the time available for the output of the domino cell to turn low when the clock falls. This delay is called the precharge delay. In general, the PMOS pull-up transistor is smaller in a domino cell than the static equivalent. Thirdly, the ratio of PMOS to NMOS transistor width for the static cell is given as 2. In Chapter 3 we will see that the actual ratios tend to be lower in static logic. Finally, for stability purposes domino cells tend to use weak feedback keepers placed between the output and the evaluation node driving the output. For simplicity, that circuit is not shown in Figure 1.4.

In addition to being able to achieve better output drive strength for input loading, domino cells also have a speed advantage as they avoid contention when the cells switch. In order to understand this, one must note that the input to a static cell drives both PMOS and NMOS transistors. Any input transition that causes the cell to switch logical states results in a PMOS transistor being turned off and an NMOS transistor being turned on, or vice versa. Since the inputs to the cell have finite rise and fall times, this means that during the transition period both the PMOS and the NMOS transistors are weakly on. This contention between the two transistors increases the input voltage level at which the cell switches. It is possible to speed up the rise or fall transition of a static cell by increasing or decreasing the ratio of the PMOS to NMOS transistor size. This, however, leads to the alternate transition becoming slower. Since both transitions are equally important in static cells, it is difficult to gain very much by skewing a particular transition. For this reason, the switching point of most static cells tends to be close to half the supply voltage level (Vdd). For domino cells only the rising transition is critical. If an input rise causes the evaluation node of the domino cell to discharge, no contention exists between PMOS and NMOS transistors. This allows domino cells to start switching when the input voltage level reaches an NMOS transistor threshold voltage level. Figure 1.5 illustrates the switching behavior of a static and a domino buffer as the data input to the cell rises. The lower switching voltage of a domino cell leads to a speedup since the input driving cells will reach the lower NMOS threshold voltage quicker than a higher voltage level. These factors lead to domino cells being significantly faster than equivalent static cells. The speed advantage of a domino cell over an equivalent static design is in the range of $1.5\times$ to $2.5\times$.

Domino logic is an uninverting style of logic [11]. This follows since every domino cell is a single-stage dynamic cell followed by an inverter. Consequently, the only valid transitions at the output of the gate during the evaluate phase are from a low to a high value. The uninverting nature of the logic means that while AND gates, OR gates, and

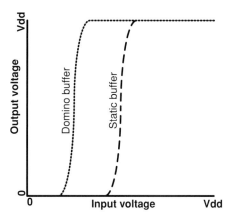

Figure 1.5. The output voltage of a static and domino buffer as the input switches from low to high.

buffers can be implemented with domino logic, NAND gates, NOR gates, and inverters cannot be. Since inverting functions are unavoidable in most designs, this would appear at first thought to preclude the general applicability of domino logic. Furthermore, as the evaluation node is precharged in domino logic cells, the only valid input transitions are from a low to a high value. Again, this is generally not acceptable. In the next paragraph we will describe how it is possible to construct general logic functions using domino logic. Before that, however, one other major difference between static and domino logic is discussed. Signal nodes, which toggle several times before reaching a final steady-state condition (referred to as glitching), are relatively common in static logic, but absent in domino designs. This is because once the output of a domino cell rises, no change in the inputs to the cell will cause the output to fall. Only when the clock falls at the end of the evaluation phase will the output of the cell fall. Thus, within a clock period there is the possibility of only one rise and fall at the output of the domino cell.

There are basically three approaches to constructing functionally correct domino logic designs in which inverting functions are present. The first approach is to stop the domino cone of logic when inverting functions are encountered. This approach was suggested in the original paper in which domino logic was introduced [11], where the XOR at the end of an adder is implemented as a static cell. The advantage of this scheme is that it allows for the easy incorporation of inverting functions with domino logic. The disadvantage with this scheme is that if an inverting function is encountered early in a path of logic, most of the gates in the path will be implemented with static logic. In addition, since some form of a latch must be placed at the boundary between domino and static logic, to ensure that the precharge value of the domino cells does not propagate through to the static logic, this will involve a timing penalty. These disadvantages could easily diminish the speed advantages to the point where it is not worthwhile. Nevertheless, if a single inverting function is present near the end of a critical path, the use of static logic at the end of the path can be a useful solution.

The other two approaches provide mechanisms to ensure that the entire path, generally from a register to a register, can be fully implemented with domino logic. The second

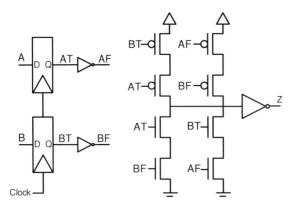

Figure 1.6. A static logic two-input XOR cell.

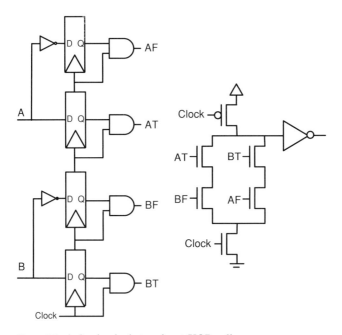

Figure 1.7. A domino logic two-input XOR cell.

approach for implementing inverting functions is to ensure that all inverting signals needed in a design are provided from the primary inputs which are low when the clock is low. In most digital designs the primary inputs to a block are the outputs of the flip-flops from a previous block. In Figure 1.6 a static logic implementation of an XOR function whose inputs are coming from a flip-flop is shown. In Figure 1.7 an implementation of the same function using domino logic is shown. To ensure that the primary inputs are initially 0, the output of the rising edge-triggered flip-flop is ANDed with the clock. This AND gate can be incorporated directly into the flip-flop. It is shown here as a separate gate for conceptual clarity. Since the clock is low before it rises, the inputs to the domino cell are initially 0. Since the XOR function needs both the inverted and uninverted versions of inputs A and B, these signals are provided directly from the flip-flops. It can be seen that the XOR function is implemented as an AND–OR gate that implements the function: $AB' + A'B$, which can then be provided to other logic. Ensuring that all the inverting logic needed by the design is provided directly from the primary inputs to the design creates a design in which no inverting functions are used. This structure is called a unate implementation. The major drawback with this approach is that it can lead to a duplication of all logic in the block, a significant area and power penalty.

The third approach to implement a correct design is to make sure that every domino cell is only clocked when stable input values are present at the input of the domino cells. For the XOR cell described above this means that if we know that after a certain amount of time the inputs are at their correct values, the clock can rise. The advantage with this approach is that inverted inputs do not need to be propagated from the primary inputs of the design. While that is a major advantage, it is difficult to do in an automated design framework. In designs generated using automated synthesis and physical design tools, inputs tend to arrive at each gate across a large window of time. Since the domino gate can be clocked only after all the inputs are stable, this means that the clock is the critical path in the gate, arriving last. This can lead to a design in which the data waits for the arrival of a clock signal at every single cell before it can proceed, slowing down the signal path. The extra margin (to account for process variation, clock tree skew, and clock driver granularity) that must be used to guarantee that the clock signal arrives after all the input signals are stable, further slows down the critical path. While difficult to use for synthesizable domino logic, it is possible to allow some binate logic for domino design implemented in custom or structured custom frameworks, where far greater control exists on the arrival time of every input. When using synthesizable domino logic in an automated flow, it is best to make the design unate. We will describe in Chapter 4 some optimizations possible to reduce the area overhead in the unating process.

One of the difficulties in deploying automated domino logic design systems is the lack of predictability, at least from a cursory human point of view, in understanding the structure of the design and the expected arrival times of signals at different inputs once they have been pushed through an ASIC flow. Gate-level netlists generated by synthesis tools appear very irregular. Layout plots highlight the differences, with one being immediately able to tell ASIC and custom designs apart. Human efforts tend to produce graphically regular, repetitive structures, while the output of an automated ASIC tool tends to appear random. Whether this phenomenon reflects any particular aptitude of human beings to discover underlying graphical patterns, or merely reflects a human preference to arbitrarily impose order, is an open question. One has to be careful also to remember that a haphazard-looking design may merely represent a degree of complexity beyond human comprehension. In the world of VLSI digital design, where we must collaborate with automated tools in all of our jobs, it is interesting to note that human designers remain in those areas where a visual or graphical, two-dimensional regularity is most present: custom, memory, and standard cell design. Where such regularity is absent, most notably implementing random logic, the engineer's task has shifted to properly operating and supervising the automated tools which do the detailed design.

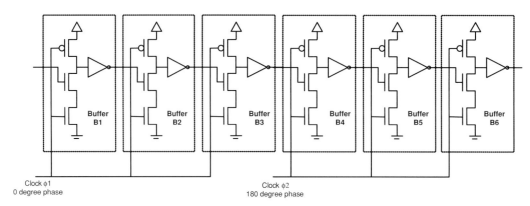

Figure 1.8. Clocking scheme for a set of domino buffers.

For RF design, perhaps the most human designer-intensive implementation task, a sharp contrast exists between the custom laid-out analog modules with their classical symmetry and palatial isolation zones, and the teeming, chaotic ASIC-designed blocks. The RF designers try to keep the digital modules with their noise spikes and unpredictable harmonics as far away as possible, although with the increasing digitalization of RF and its convergence with baseband and other purely digital functions, the haphazard cities of digital logic are continuously encroaching on the ordered world of RF.

1.3 Clocking domino logic

Ensuring that every domino cell enters the evaluate phase (when the clock is high) with its inputs low and that all inputs transition from 0 to 1 during the evaluate phase requires that the design is correctly clocked. Unlike in static logic, every domino cell is clocked. This results in the clocking strategy being far more critical and complex in domino logic, with the wrong clocking leading to possible functional failures.

The simplest mechanism to clock a domino design is to use two different clock phases. These phases can be the two clock phases of a single clock or be from two different clock sources. In Figure 1.8 a domino logic design for a series of domino buffer cells is shown. Of the six buffers, three are driven by the clock $\phi 1$, at 0 degree phase. The other three buffers are driven by the clock $\phi 2$, which has a phase shift of 180 degrees (it is inverted compared to $\phi 1$). The use of these two clock phases ensures that when buffers B1, B2, and B3 are in evaluate, i.e., data is traversing through them, the buffers B4, B5, and B6 are in precharge. Correspondingly, when buffers B4 through B6 are in evaluate mode, domino cells B1, B2, and B3 are being precharged. This is typical in domino designs where parts of a circuit are precharging while other parts are in the evaluate cycle, ensuring that as the evaluate phase data advances it encounters cells which have already been correctly precharged. This scheme ensures that the precharge delay does not impact the maximum operating speed of the digital design. Figure 1.8 also illustrates another circuit-level optimization possible with domino cells. Since all the

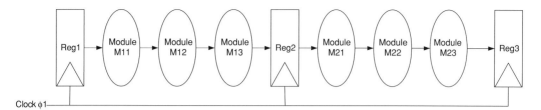

Figure 1.9. A logic module with several pipeline stages.

cells driven by a clock enter precharge at the same time, the critical path for precharge is the time taken for a single cell to precharge. The maximum evaluate phase delay, on the other hand, represents the maximum evaluate delay through a series of domino cells driven by the same clock. For our purposes this means that the sum of the evaluate delay for the buffers B1, B2, and B3 can be made equal to the precharge delay for a single buffer (B4, B5, or B6). This difference in the speed needed by the precharge and evaluate circuitry means that transistor sizing in a domino cell can be weighted to heavily favor rising output transitions over falling ones. The example shown in Figure 1.8 is simplified by not considering clock skew, cell drive strengths, and cell output loading, but it does illustrate the serial nature of the evaluate phase delay versus the parallel nature of precharge. The ratio of time allowed for precharge to evaluate depends on the number of serial cells clocked by a particular clock. This attribute is strongly dependent on the microarchitecture of the design, which determines the number of cells through which data must propagate in the evaluate phase during a clock cycle. One of the challenges in designing domino logic-compatible standard cell libraries is that one has to assume a maximum operating frequency of the design, since this sets a limit on the maximum precharge delay permissible. Setting this value too low leads to a design constrained by the precharge delays, while a very large number will optimize the precharge delay at the expense of the evaluate delay. This will negatively impact performance if, indeed, the maximum operating frequency of the design is significantly lower than that supported by the precharge delays.

Figure 1.8 shows a single stage of domino logic. When a number of different pipeline stages are present in a design, as in Figure 1.9, register stages separate the different pipeline stages. While explicit register stages are needed in static designs, to ensure that fast-arriving data does not corrupt data from an earlier cycle [12], these pipelines are not needed in domino design. Since domino cells act intrinsically as a latch, they are transparent during evaluate and shut-off during precharge, thus putting two domino gates driven by different clock phases in series causes the data to behave as if passed through a master–slave flip-flop. An alternate way to view this is to assume that in a domino design the master and slave latches present in a typical flip-flop are split and distributed throughout the logic instead of merely being at the boundary of clock phases. In addition to its faster speed, the ability to remove explicit flip-flops is one of the major advantages of domino logic, since the setup and clock-to-Q delay associated with traversing through a flip-flop can be eliminated [12].

While using two clock phases will lead to a functioning domino logic design, there are some difficulties with the approach. Care must be taken to ensure that when the clock changes phase the next domino logic cell can capture data from the previous cycle before the precharge value from the driving domino cells overwrites the valid output result. If the inputs to the domino cell on the next clock phase arrive before the domino cell is clocked, this is generally not a problem since, as discussed, the precharge delay for a domino cell is much longer than the evaluate delay through it. However, in the presence of clock skew the late arrival of the clock to the domino cell being evaluated can cause the precharge data to be clocked. This is referred to as a hold time failure. For those unfamiliar with the terminology, a quick set of definitions follows. Clock skew refers to the difference in clock arrival time between two different circuit elements which send data from one to the other either directly or via some combinational logic. For example, the clock skew between two adjacent rising edge-triggered flip-flops is the time difference between rising edges arriving at the two flip-flops. The primary cause of clock skew in digital designs is due to the presence of a clock buffering tree, a physically distributed series of buffers that sends the root clock signal to all of its nodes. For a complex ASIC with many flip-flops distributed across a large chip, some clock skew is inevitable. The other two delays associated with a clocked logic element are the setup and hold delay. The setup delay for an input to a cell is the latest input arrival time before the cell is clocked that will ensure that correct data is captured. The hold time of a flip-flop is the time after the clock edge arrives for which the input data to the element must be held at a constant value to ensure that the correct value is sampled. Setup failures are generally due to running the design too fast and can be overcome by running the design slower. Hold failures are more dangerous as they can lead to functional failures across all testing conditions and must be avoided at all costs.

The possibility of hold failure, due to clock skew, is one of the challenges of using two clock phases. In order to loosen the clock skew requirements, three or more clock phases can be used with domino logic. For this book we focus on the use of skew-tolerant clocking to clock the domino designs. This technique uses multiple overlapping clock phases [12]. Assuming that the clocks each have a 50% duty cycle, and that they are evenly distributed, each clock edge overlaps 17% with its adjacent clock. This overlap allows longer rise or fall times, reducing the clock skew requirements and hence the power dissipation in the clock tree network, while also reducing the possibility of hold failures.

There is one final advantage of domino logic that is best understood in the context of microarchitecture, where a design spans a number of pipeline stages. To understand this we look again at Figure 1.9. In Figure 1.9 a section of logic constituting a number of different pipeline stages is shown. The pipeline stage between registers Reg1 and Reg2 contains logic modules M11, M12, and M13. Logic M21, M22, and M23 is contained in the next pipeline stage. In Figure 1.10 we represent a possible domino logic implementation of the same logic. Since, as mentioned, by using multiple overlapping clock phases it is possible to replace flip-flops in domino logic, there are no explicit registers in Figure 1.10. For the design in Figure 1.9, if the first pipeline stage has the slowest delay through it, that delay will set the maximum operating frequency of the clock. This will

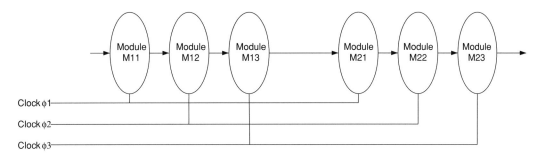

Figure 1.10. A domino logic module containing multiple pipeline stages.

not change even if the pipeline stages before and after it can be clocked much faster. For the design in Figure 1.10, a 17% overlap occurs between two adjacent clock phases. This means that if data enters module M12 even after the clock is high it can propagate through the module and can possibly use the smaller delay through M21, M22, and M23 to its advantage. This is possible because the clock for module M21 is not edge-triggered, or a hard clock, i.e., no edge-triggered value is captured at the moment that the clock rises, but rather is level-sensitive [12]. Such a clocking system is referred to as a soft or softened clock, ensuring that pipeline stages which need more time to complete can seamlessly borrow some time from contiguous pipeline stages, thus allowing for faster operating speed of the design. This property does not absolve the designer from the requirement of trying to balance pipeline stages wherever possible, or eliminate the need for synthesis and physical design tools to do the same. It does, however, provide a mechanism to equalize pipeline stage delays when granularity constraints make it difficult for different stages to have exactly identical delays. In addition, process variations due to manufacturing imperfections and cross-coupled noise may modify the delay in different pipeline stages from the assumed values. The presence of a soft clock helps distribute these silicon variations, limiting the impact on operating frequency. While this technique can be, and indeed is sometimes, used with static logic, it comes for free with skew-tolerant domino.

1.4 Summary

The use of domino logic, like just about any other design choice, has its advantages and disadvantages. The primary advantage of domino logic is its speed of operation. The advantage comes from the superior speed of the domino cell and the ease with which explicit flip-flops and clock softening can be supported with domino. The primary disadvantage of domino is its greater power dissipation, its potential area penalty, and the complexity of using domino logic.

The higher power dissipation in domino logic is due to the fact that every single domino cell is clocked. Power must hence be consumed to switch the domino cell every single clock cycle. This power dissipation occurs irrespective of whether the data is traversing the domino design or not. In addition to this, domino logic cells need to be precharged if the output is a logic one value at the end of the evaluate phase. This leads to an extra transition that static logic does not require. If we construct a domino logic circuit by providing the true and false values for every input, there is a possible doubling of the number of logic gates and a proportional rise in its power dissipation.

These factors suggest that a synthesizable domino flow generally is not useful for low power modules (in custom design great effort can be used to minimize the power impact of domino). The easiest way to limit the power overhead is to limit its use to those areas in a system-on-chip (SOC) where the requirements for speed necessitate its use. For an SOC the tradeoff in power between domino and static logic is more complex than at the cell or module level, since system-level considerations come into play. For example, if the use of domino logic allows two cores to be replaced by a single core running much faster, the overall power penalty for domino logic is reduced. Alternately, if domino logic allows a timing-critical circuit to be used with low leakage transistors, while a static implementation requires faster but leakier transistors, the power difference is more difficult to predict. Aside from the system considerations, at the circuit level there is one power advantage that domino designs have: their absence of glitching. Glitching leads to extra power dissipation, due to logically unnecessary switching being absent in domino.

The area overhead of domino designs is again somewhat difficult to generalize. In custom applications, domino designs can be smaller than corresponding static designs. For synthesizable domino logic, the extra logic can lead to a power penalty. Since it is assumed that synthesizable domino logic will be used sparingly, this overhead may be acceptable for most designs.

The last and final disadvantage of domino is the complexity of the flow. Using synthesizable domino logic hopefully manages to hide the details of the complexity from the users. This will also reduce the great time and effort required by custom designs, although the design will be less optimized than a custom design.

Despite these disadvantages, the expeditious use of domino logic does have some advantages. In addition to the system-level advantages mentioned earlier, there are always some logic modules where faster speed can be utilized for greater advantage. Much of the current research on analog design focuses on the use of very high speed digital to complement and control analog circuitry. Again this represents a small, but critically important, part of the total circuitry in the design, where domino can be used to its speed advantages. Another possible application of synthesizable domino logic is in speeding up a legacy module without incurring the cost of redesigning the microarchitecture and porting the software. With a synthesizable domino logic solution, the effort involved in using it becomes much lower, making it much more attractive beyond its traditional application areas of high end microprocessors.

References

1. F. Wanlass *et al.*, Nanowatt logic using field-effect metal-oxide semiconductor triodes, International Symposium on Solid-State Circuits, 1963.
2. http://www.icknowledge.com/history/1960s.html [accessed 29 June 2007].

3. http://en.wikipedia.org/wiki/CMOS [accessed 29 June 2007].
4. http://www.icknowledge.com/history/1970s.html [accessed 29 June 2007].
5. K. Bernstein, 'Out-of-the-park home runs', legendary digital circuits that tracked technology scaling, *IEEE SSCS Newsletter*, Spring 2007.
6. S. Wolf, *Silicon Processing for the VLSI Era*, Volume 2: *Process Integration*, Lattice Press, Sunset Beach, CA, 1990.
7. A. Sangiovanni-Vincentelli, The tides of EDA, *IEEE Design and Test of Computers*, November–December 2003.
8. R. Goering, EDAC: EDA up 15 percent in 2006, http://www.eetimes.com/news/latest/showArticle.jhtml?articleID=198900043
9. http://i.cmpnet.com/eetimes/eedesign/2007/chart1˙031507.gif
10. I. Sutherland, B. Sproull and D. Harris, *Logical Effort: Designing Fast CMOS Circuits*, Morgan Kaufmann Publishers, San Francisco, CA, 1999.
11. R. H. Krambeck, C. M. Lee and H-F. S. Law, High speed compact circuits with CMOS, *IEEE Journal of Solid-State Circuits* **SC-17**(3), June 1982.
12. D. Harris, *Skew-Tolerant Circuit Design*, Morgan Kaufmann Publishers, San Francisco, CA, 2001.

2 High-speed digital design

2.1 Microprocessors since 1989

In 1989 a forward-looking paper attempted to determine the characteristics of microprocessors in the year 2000. Called "Microprocessors circa 2000", the paper hypothesized that a high-performance microprocessor in the year 2000 would have an area of 1 square inch (645 sq mm), contain 50 million transistors, and run at above 250 MHz [1]. The overall performance of the microprocessor was estimated at 2000 million instructions per second (MIPS), achieved by the employment of two or three cores, each with a performance of 750 MIPS. Forward-looking papers often have somewhat fanciful conceits of future developments, illustrating the witticism that predictions tend to be difficult if they involve the future. This prediction, however, was based on many years of microprocessor development, leading to a broadly accurate prediction of things to come. The International Solid State Circuit Conference (ISSCC), held in early 2000, presented a number of microprocessors whose transistor counts and area were within $2\times$ of the prediction. Since much of the area of a microprocessor is composed of on-chip memory, the prediction for transistor count was achieved soon afterwards. The prediction of 2000 MIPS for the maximum performance of the system also proved to be accurate. The interesting discrepancy was in the way that the performance of the microprocessor was achieved. Instead of employing a number of processors operating at 250 MHz, most high end microprocessors were single core designs running at or above 1 GHz. At the ISSCC held in early 2001, the Pentium 4 microprocessor was introduced by Intel. The clock frequency of this processor was 2 GHz, with the integer execution unit running at 4 GHz.

How had such a historically anomalous jump in clock frequency been achieved in such a short period of time? Much of this can be attributed to the intense competition in the microprocessor market, especially for processors compatible with Intel's $\times 86$ family of chips. For many customers a very high clock frequency had become a critical component in customers' expectations when buying a computer. It can be debated how important clock frequency is for a computer, especially since memory and I/Os are often more severe bottlenecks. Still, clock frequency was a major marketing advantage throughout this period, although, of late, its centrality has started to wane somewhat. In addition, in 1992 Digital Computer Corporation released the first version of its Alpha processor, which achieved an impressive clock rate of 200 MHz. Since 250 MHz was the expected clock rate at the end of the decade, Digital's ability to reach comparable speeds eight years early spurred clock speeds to be pushed throughout the industry.

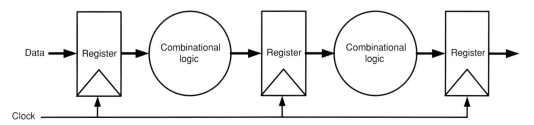

Figure 2.1. Pipeline stages with combinational logic between the registers.

While competition can explain the motivation for desiring high operating speed, it does not answer the basic engineering question as to how this was achieved? The answer lies with improvements in CMOS processes and the use of much more aggressive designs, specifically, the use of much deeper register pipelines.

CMOS transistors in digital designs can be viewed simplistically as switches. As new processes with smaller feature sizes become available, it is possible to scale the physical dimensions of the transistors. For digital designs this leads to many advantages. Smaller transistors require less energy to turn on and have smaller capacitive self-loading. They are also faster. The speed, power, and area benefits due to shrinking a process can be derived analytically [2]. Let us assume that at each new process generation the lateral and vertical dimensions of the transistor are scaled by $1/S$, where S is called the scaling factor. Over the last few process generations the typical scaling factor has been approximately 1.4. Since supply voltages and gate oxides are typically also reduced, they are also assumed to be scaled by $1/S$. This leads to the delay in the transistor being reduced by $1/S$, power dissipation scaled by $1/S^2$, and a power density that remains constant. Scaling leads to a reduction in area of $1/S^2$. This means that a significant reduction in silicon area, or a large increase in functionality, is possible by porting a design to a new CMOS process. Both of these factors help to reduce total system cost, which drives demand further; providing positive feedback for the virtuous cycle of demand, supply, and engineering employment.

The other factor mentioned which led to improvement in microprocessor clock frequencies through the 1990s was much better implementations, spearheaded by the use of aggressive microarchitectures. These microarchitectures were much more deeply pipelined. Figure 2.1 shows a pipeline stage with registers and combinational logic. This combinational logic is called a cloud of logic and implements the function of the circuit (due to the lack of an appropriate drawing stencil the clouds in Figure 2.1 are rendered as somewhat circular in shape). The total delay for a pipeline stage is the delay through the combinational logic between the pipeline stage and the clocking overhead. The clocking overhead corresponds to the clock-to-Q delay at the flip-flop launching the data, the setup time at the flip-flop capturing the data, and the clock uncertainty due to skew and jitter. Increasing pipeline depth reduces the logic stages between flip-flops. As we discussed in Chapter 1, it is possible to reduce significantly clocking overhead with the use of domino logic and overlapping clock phases [3]. For deeply pipelined designs this is particularly important due to the large proportion of total delay

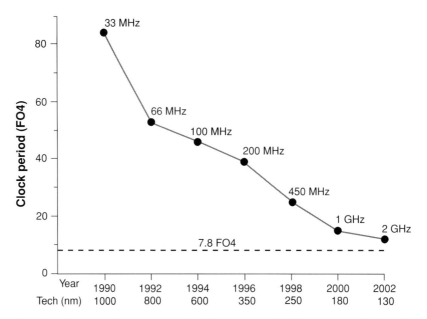

Figure 2.2. The year of introduction, clock frequency, and fabrication technologies of the last seven generations of Intel processors.
Source: Used with permission from the paper "The optimal depth per pipeline stage is 6 to 8 FO4 inverter delays" by M. S. Hrishikesh et al., International Symposium on Computer Architecture, May 2002.

consumed by clocking overhead [3]. For Intel processors in the 1990s there was a large reduction in the logic between adjacent flip-flops, indicating the increasing emphasis placed on deeply pipelined microarchitectures [4]. This is shown in Figure 2.2, where the delay is given as the equivalent number of inverters driving four other inverters (FO4). Since this measure considers delay in terms of inverter delays and not absolute time, it is a useful metric to compare microarchitectures in a technology-independent manner.

Figure 2.2 shows that between 1990 and 2000 the total number of FO4 stages went from 80 to 10, an approximately 8× reduction. Assuming that a typical CMOS process generation leads to a delay reduction of 1.4× compared with the previous generation, an 8× improvement in delay corresponds to the delay reduction achieved by 4.4 process shrinks. This is a remarkable figure. Developing a new process node is extremely expensive due to the high development costs and capital expenditure. Achieving speed improvement equivalent to multiple process shrinks from improved design has tremendous benefits not only from a cost point of view, but also by allowing far more capable designs to come to the market earlier. While Figure 2.2 shows a reduction in logic between registers, microarchitectural improvements entail more than merely placing more registers in the design. Efficient design requires that a certain granularity of tasks be performed during each clock cycle. In order to ensure that this could be achieved with fewer FO4 inverter delays of logic, domino and other custom design methodologies started to be used extensively for very high-speed designs.

While the use of much deeper pipelines has led to greatly improved performance in microprocessors, there are limits to the extent to which this approach can be applied. Figure 2.2 is borrowed from a paper entitled "The optimal logic depth per pipeline stage is 6 to 8 FO4 inverter delays", which suggests that the current processors are beginning to approach the limits of maximum pipelining [4]. Further pipelining, which leads to more instructions being processed at the same time, does not help overall performance since the likely occurrence of a jump instruction will lead to the pipeline needing to be flushed. Synthesizable processor cores implemented with automatic ASIC design flows are mimicking many of the microarchitectural features found in custom designs, including the use of deep pipelines [5]. Synthesizable logic does not, however, have access to domino logic or other dynamic logic styles. This is a major difference between custom and ASIC design methodologies [5].

A point to note about high-speed ASIC design modules is to remember that while processor design remains one of the most studied and perhaps "glamorous" aspects of digital implementation, it is not necessarily representative of ASIC designs. The ASIC market and the high end microprocessor market are fundamentally different in terms of both the application space and the economies of the business. Examples of large digital ASICs include graphics processors, cell phone baseband processors, and set top box decoders. The critical computational modules in these blocks are often data-processing functions which need to be implemented under very strict power and cost budgets. Data-processing functions also tend to possess considerable intrinsic parallelism. These functions are most competitively implemented in hardware and not with a programmable microprocessor.

Since the end markets for the chips are in the consumer space, the average selling price (ASP) for these products falls sharply with time. From a purely business point of view this means that it is essential to reach the consumer markets with products quickly. For this reason semiconductor companies are extremely uneager to use custom solutions, such as domino logic, in ASIC products. Since high end microprocessors typically command much higher prices, they strive to provide performance as a primary design requirement. This leads to relatively long design cycles and almost mandates the use of considerable customization. In addition to the faster time-to-market requirement, the applications targeted by chips implemented with ASIC methodologies are much more cost-sensitive. This should be no surprise to anyone who has purchased a DVD player and a personal computer of late, since the prices paid by consumers are ultimately reflected in what semiconductor companies can charge for their components. I remember once asking a colleague in sales about his best business deal. The only chip pricing I had been familiar with had been microprocessors, which were then available for prices in the range of several hundred dollars each. My colleague thought for a second, until a large smile broke across his face and he recounted the sale of a highly profitable part that he had sold for less than three dollars a piece. At the time I was somewhat surprised by the number, it seemed so low. As electronics is increasingly directed towards the consumer market, it is worthwhile remembering that a few dollars remains the ASP for a large and ever-increasing set of chips. Like supermarkets, the real money is made in volume (even though the ASPs for semiconductors are low, the overall profit margins for the industry are much higher than those for supermarkets). It is in this context of falling ASPs and rapid time-to-market pressures in the ASIC design space that a synthesizable domino logic solution is proposed in this book.

Since domino logic will only be applied if existing static logic-based synthesis has not achieved timing closure, the next section deals with techniques used to improve the implementation of ASIC designs. Many of these techniques represent good design practice and can be applied whether the underlying circuit technology uses static or domino logic.

2.2 Microarchitectures for high speed

According to IBM's original definition for computers, architecture refers to the definition of the instruction set and other programming abstracts. The actual act of transferring this instruction set to hardware was referred to as organization [6]. In this book we will consider microarchitecture or logic design to refer to the design of all the logic. We will adhere to the convention of referring to architecture as meaning the software and algorithmic view of the design. The microarchitecture is the next step of the design, where the architecture is mapped to logic and the first realistic timing and area estimates made.

As mentioned earlier in the chapter, high clock speed requires the use of deeply pipelined designs. Inserting extra pipeline stages does not of itself guarantee high performance, since it may be difficult to partition the logic evenly across a number of different stages. Furthermore, RTL code that appears to be evenly balanced may turn out to be less so once implemented. Under these conditions, considerable effort must be expended to ensure that the design can run with a fast clock. Such efforts are the bread and butter, or perhaps more appropriately in this age of globalization, the rice and dal, of digital design. While it is possible to enumerate different approaches to achieving an efficient microarchitecture, it may be more instructive to study some examples.

2.2.1 Fast arithmetic modules

Critical paths, or the longest delay paths in digital designs, often traverse through fast adders, multipliers, and other arithmetic functions. Sometimes these paths cannot be further pipelined, since they have feedback requirements which require the full computational cycle to be completed every clock period. Examples of such circuits include accumulators and Viterbi add–compare–select modules, in which data must come out of a register, go through some logic, before being latched in the same register at the end of the clock cycle.

There has been much research on designing high-speed arithmetic modules [6]. Rather than repeating well-known techniques for speeding up these operations, an example circuit is discussed to give the reader an idea of the kind of techniques used. A general principle in speeding up arithmetic operations is to parallelize the operation as much as possible. A ripple carry adder, for example, has a delay that is proportional to the

High-speed digital design 23

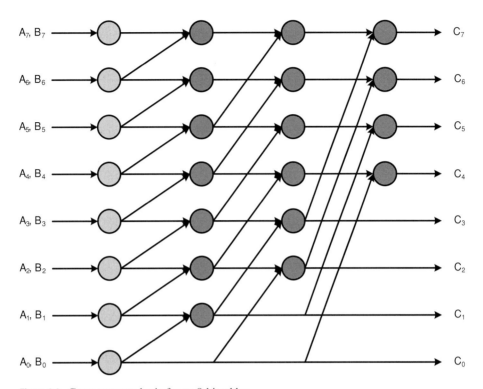

Figure 2.3. Carry generate logic for an 8-bit adder.

number of input bits. While area-efficient, the ripple carry adder is slow, especially at large sizes. The simplest way to parallelize an adder is to make every single output a direct function of its cone of dependent inputs. This quickly leads to an unrealizable solution, since the most significant bit (MSB) output of a 32-bit adder is dependent on all 64 input bits (32 bits from each input bus). The solution is inelegant not only due to the massive area and extensive fan-out requirements for the inputs (these two factors will counteract the very low logic depth of the design and slow it down considerably), but also because considerable replication is present at each output node. A better solution can be achieved by applying the Kogge–Stone algorithm [7]. The Kogge–Stone algorithm describes a method by which certain recursive functions can be parallelized. Since the addition operation can be defined recursively, in terms of the carry propagate logic, it is amenable to this approach.

In Figure 2.3 the carry propagate logic for an 8-bit Kogge–Stone adder is shown. The reader is reminded that the output for each bit of the adder is the exclusive OR (XOR) of the two input bits and the carry-in for that bit. The lighter gray circles in the figure implement the bit-level carry generate and propagate function. These functions are: A_i & B_i for generate and $A_i \mid B_i$ for propagate, where A_i and B_i are the input bits, & is the logical AND function, and | the logical OR. The carry-in bits then progress to the darker circles, which implement the next carry generates and propagates. These functions are defined as: $C_{i+1} \mid C_i$ & P_{i+1} for generate, and P_{i+1} & P_i for propagate. The variables

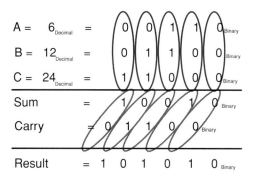

Figure 2.4. Adding three numbers to produce a carry and save term with the final result having the correct value.

C_{i+1}, P_{i+1}, C_i, and P_i refer to the higher- and lower-order carry and propagate inputs to the cell.

The structure of the schematic shown in Figure 2.3, where the inputs to the logic at every stage span across progressively greater widths, is common in many high-speed arithmetic modules. The total number of cells in the critical path for the design is proportional to the base-2 logarithm of the number of inputs to the adder. Shifters, in which data can be shifted left or right, tend to look similar, as do increment functions and priority encoders. It is possible to modify the tree structure shown by having each cell be fed from more than two inputs. In such implementations the logic depth of the tree decreases, although each node has more logic. The optimality of a particular design depends on the tradeoff between fewer but more complex logic functions versus faster but more numerous elements. Much of this is technology- and circuit design-specific, especially since it must also consider the delay associated with the wires used to route the signals.

The reader may have noticed that the set of arithmetic functions mentioned earlier did not include multipliers, another common arithmetic function. Indeed, the implementation of a multiplier is a bit different since it requires two separate functions to be performed. By way of explanation, we note that a straightforward approach to perform multiplication is through repeated addition. This can lead to very long computation times since multiple additions are needed. To avoid this delay, the multiplier performs an initial carry save compression. At the end of the carry save compression the generated terms are added with an adder. The final adder is also called a carry propagate adder, since the carry can propagate across the bit width of the adder. In order to illustrate the difference between carry save and carry propagate addition, a simple example is shown in Figure 2.4. Three numbers 6, 12, and 24 are represented in their binary formats. Since we do not wish to use two separate adders to add the three numbers, the three terms must be reduced to two and then added. The three terms can be compressed by using a full adder. A full adder receives three inputs and produces two outputs. The sum output is the exclusive OR (XOR) function of the three column inputs. The sum output has the same binary position as the inputs to the full adder. The carry output is set to one if two or more inputs are one. This function is also called a majority voter since the output bit reflects which input bit is in the majority. The sum output is shifted one binary position to the left of the input bits. The full adder works on bits from the same binary position in the numbers being added together. In Figure 2.4 these numbers are circled. The resultant carry and sum buses are shown below. The outputs of each full adder are shown within the slanting ovals. Once the data is in the form of a sum and carry bus, it is added in a conventional adder. The full adder in this example acts as a carry save adder since all it does is to help compress three terms to two. The carry save name comes from the fact that every output is a function of a number of input bits, with no carry propagating across cells. The delay of a carry save adder is, thus, a single-cell delay, while a carry propagate adder will need to traverse through a number of cells.

Most multipliers will have more than three input terms. This requires a number of full adder stages to compress the data to two final carry save terms, which can then be added in an adder. In addition to full adders a number of different compressor circuits can be used, such as a four-term to two-term carry save adder (CSA42) and a five-term to three-term carry save adder (CSA53). These compressors have a higher compression ratio than a full adder (a full adder compresses three inputs to two outputs, whereas a CSA42 compresses four inputs to two outputs). They are, however, larger and hence slower cells. The optimal configuration involves a tradeoff similar to that involved in designing a fast adder: choosing between a fewer number of more complex cells, or a larger number of faster ones. The best solution depends on the delay of the different compressor cells, the multiplier size, and the routing delay. These metrics can vary from process to process. In addition to multiplication, carry save addition can be used to implement three or more input adders.

While understanding the approaches used to perform high-speed arithmetic operations is often useful, such knowledge is not essential in ASIC design. Synthesis tools are increasingly knowledgeable about different algorithms for implementing arithmetic functions. Tools such as Synopsys's Design Compiler^{TM} are also capable of optimizing datapath logic across different modules. For example, a number of sequential adders placed in the RTL will automatically be replaced by the more optimal solution of using carry save addition and only one final carry propagate adder. While it is possible sometimes to improve the performance of a design by directly coding an optimized logical solution in an RTL language, such as Verilog, such efforts tend to have a limited utility for a great deal of effort. It is probably best to use these techniques only when a new non-standard arithmetic module is needed, or when custom standard cells are available. The task of implementing arithmetic functions in ASIC design flows is increasingly best left to computers and software.

2.2.2 Predictive logic and parallel computation

While competing against synthesis tools in implementing well-understood arithmetic modules is not recommended by the author, it does not exclude intervention by the designer to achieve the required performance when direct synthesis is proving inadequate. In order to achieve the desired results a set of optimizations can be attempted, using techniques that often change the abstraction level or the problem formulation. Hopefully some examples will help fill in the wonderful vagueness of what is meant by that!

A seemingly intractable critical path emerged during the design of a deeply pipelined microprocessor that the author was involved with. The function implemented by the logic was to determine if a jump in the instruction sequences was needed. Jumps or conditional branches are common instructions, representing approximately 20% of executed instruction [6]. The actual task involved the result of a comparison following an addition. A direct RTL description of the required function did not meet the target frequency of the processor.

Algebraically formulating the problem led it to be defined as the addition of two integers X and Y, followed by a check to see if their sum is equal to integer Z. For our consideration X, Y, and Z are assumed to be 32-bit numbers, coded in standard 2s complement format. The analysis can easily be extended to different bit widths. Thus, we have to check if: $X + Y = Z$. This is equivalent to $X + Y - Z = 0$ (for the punctilious this follows from the presence of the additive inverse in boolean algebra). Since a negative number is defined by bitwise negation and the addition of a 1 in 2s complement format, the problem could be expressed as: $X + Y + \sim(Z) + 1 = 0$. The only way that adding a 1 to a binary number leads the output to be 0 (within its bit width) is if the number that the one is being added to is all ones. This follows since the one will cause a final carry-out bit to be generated beyond the specified bit width, with all bits in the specified range being zero. Thus, for our considerations, determining if the result of the addition and comparison is equal to zero is equivalent to determining if: $X + Y + \sim(Z) = 32'hffffffff$, where 32'hffffffff is the Verilog representation of 32 ones in hexadecimal form.

The reformulated solution does not of itself lead to a better solution, but it does provide insight into how it can be done. When the sum of a set of numbers has to be equal to a particular value, it is sometimes possible to determine this by bitwise operations followed by a check to make sure that all the bitwise results are correct. For our purposes this involved ensuring that at every bit position the sum of X, Y, $\sim$Z, and the carry-in bit from the previous bit column is equal to 1. This is done by using XOR logic. The carry-in bit is calculated by checking to see if at least two bits in the previous bit column are equal to 1. The final result is calculated by ANDing the result for all bits. While such a scheme may seem to be somewhat suspect as a general approach, it can be formally shown that the result of this operation is always correct!

In Figure 2.5 a simple example with 4-bit operands is used to illustrate this approach. For the example on the left of Figure 2.5, the sum of $+7$ is added to -2 and compared with the expected result of $+5$. For the example on the right a different number is used. The comparison result is met when all the output bit functions are 1. This can be checked with AND logic. The delay for this approach, a bit operation followed by a large AND gate, is significantly less than performing a full add followed by a comparison. To compare the delay in both approaches, a straightforward implementation and the proposed technique were synthesized using a 0.18 μm CMOS library. Synthesis results showed that the conventional approach required 2.25 ns, while the proposed implementation required 1.27 ns. This corresponds to a 44% reduction in delay [8]. The proposed solution is also smaller and less power-hungry.

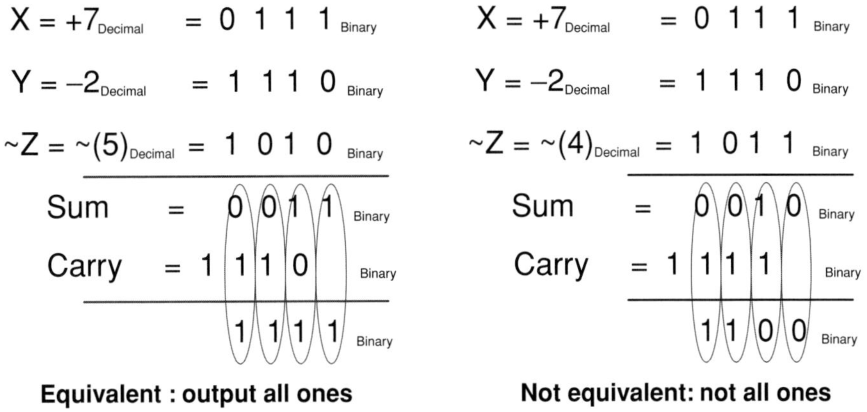

High-speed digital design

$X = +7_{Decimal}$ $= 0\ 1\ 1\ 1_{Binary}$ $X = +7_{Decimal}$ $= 0\ 1\ 1\ 1_{Binary}$

$Y = -2_{Decimal}$ $= 1\ 1\ 1\ 0_{Binary}$ $Y = -2_{Decimal}$ $= 1\ 1\ 1\ 0_{Binary}$

$\sim Z = \sim(5)_{Decimal} = 1\ 0\ 1\ 0_{Binary}$ $\sim Z = \sim(4)_{Decimal} = 1\ 0\ 1\ 1_{Binary}$

Sum $= 0\ 0\ 1\ 1_{Binary}$ Sum $= 0\ 0\ 1\ 0_{Binary}$

Carry $= 1\ 1\ 1\ 0_{Binary}$ Carry $= 1\ 1\ 1\ 1_{Binary}$

$1\ 1\ 1\ 1_{Binary}$ $1\ 1\ 0\ 0_{Binary}$

Equivalent : output all ones   **Not equivalent: not all ones**

Figure 2.5. Numerical example of a predictive adder and comparator.

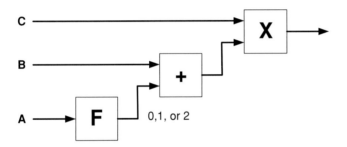

Figure 2.6. Serial computational function with a multiplication at the end.

Another technique that can be used to speed up a function is applicable if it consists of a number of serial functions, which can then be replaced by a set of speculative parallel operations. The parallel operations compute the set of possible outcomes for the original operation, with logic operating in parallel determining which outcome to select. To illustrate this situation, let us assume that A, B, and C are provided to a module. Depending on the value of A, B, and C, B + 1 and C, or B + 2 and C, are multiplied together. Schematically this is shown in Figure 2.6, where the module F computes whether 0, 1, or 2 is to be added to B before being multiplied by C.

If the function shown in the Figure 2.6 operation cannot be completed in a clock cycle, it is possible to parallelize the addition and multiplication steps (A and B) with the selection logic F. This is shown in Figure 2.7. Here it can be seen that all three possible multiplications start earlier. This is useful since the multiplication will likely have the longest delay of all the functions. In parallel, the function F computes which of the outputs to select. This is done via a multiplexer, shown as module M. Of course, this approach is worthwhile only if the delay for module F is much greater than the delay through the multiplexer. This technique is quite useful and tends to pop up in circuit and architecture design as well. By speculatively computing operations at the same time as the selection logic, the delay of the selection logic can often largely be masked by that of

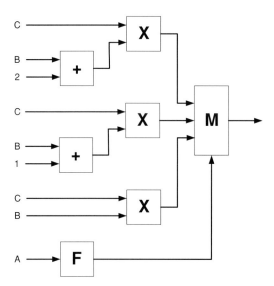

Figure 2.7. Parallel implementation of the function shown in Figure 2.6 with the multiplication starting earlier.

the other operations. Although costly in terms of area and power, this can significantly reduce delay. As of now, logic synthesis tools do not automatically invoke this form of structure from a straightforward RTL description. One of the reasons is that this approach has extra functionality, or logic redundancies, which tends to run counter to the goal of synthesis tools, which strive for non-redundant implementations.

A practical circuit where this approach has been used is in the design of a floating point unit (FPU). Floating point arithmetic is used in calculations requiring decimal precision, such as scientific calculations. The standard representation of a floating point number is as a fractional part and an exponential value. For example, in decimal logic 625.7 is expressed as 0.6257×10^3, where 0.6257 is the fractional component of the number and 10^3 the exponential power. When two fractional numbers are to be added or subtracted from each other, the number with smaller exponential component must be right-shifted so that the representation of the fractional parts corresponds to the same exponent base. Thus, if 20.0 or 0.20×10^2 is subtracted from 625.7, 20.0 should be expressed as 0.020×10^3 to allow the fractional components to be subtracted directly. If floating point addition or subtraction is to be performed on two numbers, their exponents are first subtracted from each other. This determines which, if any, number has a larger exponent. Subsequently, the number with the smaller exponent will need to be shifted by the difference in the exponent values. Depending on which exponent value is chosen first, the result of the exponent difference can be positive or negative. If it is negative it needs to be converted back to a positive number by bitwise negation and adding a 1 to it. Converting the difference in exponents to a positive value leads to an extra delay [9]. This can be avoided if the two exponents are simultaneously subtracted from each other, with the result of the subtraction used to shift the two floating point fractions. The result is generally only valid for one of two values on which the subtraction and shifting

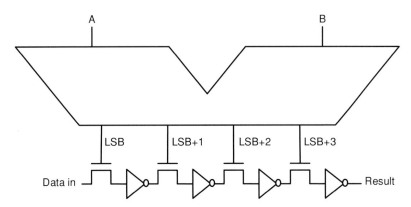

Figure 2.8. An integrated adder shifter design.

is occurring [9]. Having both operations proceed from the start, however, ensures that the extra delay of converting a negative exponent to a positive one does not have to be incurred. Selecting between the two operations can be left to later, by which time the difference in the exponent values is known. The only speed penalty with this approach is the need for an extra multiplexing function to choose the correct result. Since multiplexers were already present in the design, the extra multiplexer required only a small penalty.

2.2.3 Optimizing across logic and circuit design

One of the reasons that custom design can achieve results superior to automated flows is that it easily allows optimizations across different design abstractions in a flow. A microarchitectural roadblock that may be intractable at the logic level, for example, can be made more manageable by a change in the architecture. Similarly, by allowing designers the freedom to design both the logic and circuit transistors, it is often possible to achieve better results than sharply compartmentalizing the two. Of course, to be effective this requires designers to be familiar with different levels of design abstraction. This is often difficult for engineers to acquire, due to the rigid separation of tasks in engineering organizations (for engineering managers reading this I urge them to allow a little more flexibility, wherever possible, in their organizations).

To see how logic and circuit design can be designed in a complementary manner, we shall look at an adder shifter block. This particular design was done in a 0.35 μm CMOS technology, in which the shifter was implemented using NMOS pass transistor logic and inverters [9]. A schematic of the path for a single input data is shown in Figure 2.8. At the beginning of the clock cycle, inputs A and B are provided to the adder along with "Data in", which is a bit in the fractional component of the floating point number. In order to speed up the operation of the adder shifter block, initial design efforts focused on using a fast adder. Since a Kogge–Stone adder was available, it was chosen. It was discovered, however, that using a fast parallel adder did not lead to the fastest module performance. In order to explain this seemingly paradoxical situation, the reader is referred to Figure 2.8. It can be seen that until the LSB output of the adder is available, data does not start to propagate through the shifter. While having a fast, parallel adder architecture leads to the fastest adder, this speed is measured from any input to any output, which for adders typically means from the LSB or another lower-order bit, to the MSB or a high-order bit. The overhead of the Kogge–Stone adder, however, meant that the output of lower-order bits was slower than when a simpler adder architecture, such as a ripple carry, was used [10]. The final architecture for the adder was a mixture of ripple carry for the lowest bits and a more conventional faster architecture for the higher bits.

The example of the adder shifter design was in the context of a structured custom design. For ASIC design flows the best circuit design improvements can generally be made by observing how the delay of the critical path can be reduced by adding specific standard cells to the library. These cells, sometimes called hot cells, can then be designed and included in the library. The benefits from using these hot cells are often design-specific. For best performance the custom cells may need to be pre-instantiated in the design. This is most often the case when the cell is a flip-flop with built-in logic or a very complex standard cell. It is somewhat perplexing that providing the custom cells to the synthesis tools can still lead to an inferior solution than achieved with hand instantiation. This is most noticeable when a particular logic structure of limited cell depth is needed to meet timing. It may be that a particular hand-instantiated solution represents a global optimum which, depending on the technology mapping step during logic synthesis, becomes unreachable for a synthesized solution. In addition, technology mapping algorithms are often simplified to ensure reasonable runtime during logic synthesis, limiting the quality of the solution. While it may be encouraging to think that these cases suggest the superiority of human designers over synthesis tools, one has to remember that analyzing and adding the custom cells occurs only after the synthesis tool has optimized the design extensively. The work of the human designer is thus more collaborative than directly competitive with the synthesis tool.

2.2.4 Remarks

One of my professors in graduate school was deeply enamored of giving questions in tests which could only be correctly completed in a timely manner by the application of a specific trick. I have never been a big fan of puzzles, and can share the reader's concern about the rather haphazard nature of the problems raised and solved in the preceding pages. Facing the choice between being comprehensive but vague, or specific if arbitrary, I chose the latter, more practical approach. The engineers' approach, if you will.

The discussions till now have focused almost exclusively on high-speed design. This has been done in the context of assuming that the architecture has been properly chosen. For example, it may be possible that some form of parallelism can allow a design to achieve performance without stressing the speed of the design. Most ASIC engineers do not, however, decide the architecture of the design. This is provided. Performance, defined by the speed of the circuit, must then be met. There has also been little discussion of power and area. For very high-speed designs, power and area must almost by definition be subservient to the needs of speed. Synthesis tools will attempt to use the slack in non-critical paths to reduce the area and the power of the design wherever possible. Often it is possible to choose between area and power as the most important consideration after meeting timing. It has been suggested that the best way to achieve low power in a design is to push for maximum speed, beyond specified needs, and then reduce speed by lowering voltage [11]. While promising, this technique is often difficult to realize due to designers being constrained in using only specific power supply voltages.

2.3 Designing and using high-speed memories

Memories, such as static random access memories (SRAMs), read only memories (ROMs), and register files, tend to heavily populate ASIC designs. Since these memories often fall along the critical path in the design, a short overview of memory design is presented here. The relative speed of memory versus logic often defines many architectural choices. For example, the use of complex instruction set architectures in early computer systems was based on the use of off-chip instruction memories in the multi-chip systems then present. Deeply encoded instructions were preferable since the long decode times could be masked by the chip-to-chip communication delays. Only when semiconductor manufacturing technology improved to the point where chips with on-chip memories became possible was it worthwhile to move to reduced instruction set computing (RISC) architectures. In RISC machines, simpler and faster decode logic was necessary to keep up with the faster memory access times [11].

A simple example of a 4×4 SRAM memory is shown in Figure 2.9. The components of the memory are: the row decoders on the left, the memory cell array in the center, the input data buffers at the top, and the sense amplifiers at the bottom. Figure 2.9 gives an overview of all the major components in an SRAM. The decoder logic ensures that after decoding, a single row of memory addresses are enabled. The memory array stores the memory value as horizontal data words. The sense amplifiers are used to read data from the memory array, while the input buffers are used to drive the input data into the memory array.

Most ASIC designers receive memory modules from internal or external memory design groups. At the logical level a memory can be considered to be similar to a flip-flop, in that the memory has a setup time, a clock-to-Q delay, and a clock insertion delay. The setup time of a memory is the time needed for the data to be written to the memory array. If the decoded signal address arrives with the data, the critical path for the setup delay will include the time needed to decode it. If, however, the address comes earlier than the data, the address path can be enabled allowing for a faster write. As in flip-flops, it is possible to change the setup and clock-to-Q delay for a memory by having more or less delay on the clock insertion delay. This may be useful for cases in which slack exists on the pipeline stage preceding or following the memory module. Techniques in which clock delays are modified to favor one side of a pipeline stage represent zero-sum situations, i.e., the extra time available to compute a function on one side of the pipeline is offset by having correspondingly less time on the other side of the

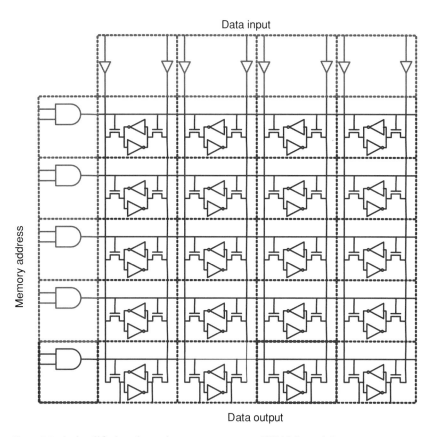

Figure 2.9. A simplified static random access memory (SRAM) module.

pipeline. These methods are, hence, only helpful when slack exists on one side of the memory boundary.

While a straightforward clocked memory module is described here as a single pipeline stage, memories need not correspond to this. A memory such as a ROM, from which data is read and subsequently used, can be constructed to be a purely combinational logic function, without any flip-flop boundaries. In such designs the memory tends to be asynchronous or controlled via some enable signal. Alternately, it is possible to pipeline the memory more deeply. Clock boundaries may be placed in decoding logic or sense amplifiers in addition to that present in the memory core. There are, however, limits to how much pipelining is practically possible within the memory array, since much of its functionality is analog in nature. One of the ways to build faster memories is to break up long bit lines into smaller segments. These local, less heavily loaded bit lines can then be sensed with local sense amplifiers.

If the speed of the memory provided is not adequate or latency constraints do not allow the use of a more deeply pipelined memory, then it may be possible to speed up the memory by recognizing some design-specific constraints. For example, for the case already discussed, where the data address for a memory comes early, it is possible to

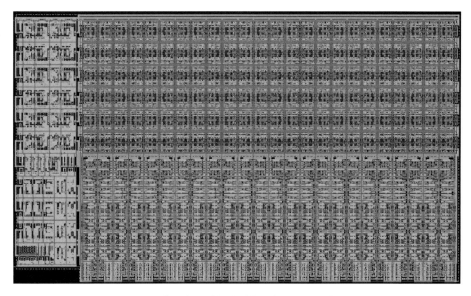

Figure 2.10. Layout plot of a small register file (24 bits wide and 12 bits deep).

achieve speed with a more accurate characterization of the memory. Alternately, if a particular bit in a memory is critical, it may be possible to favor the access time for that bit or store it in a register, making sure to correctly use that value when it is needed. When the whole memory speed is lagging despite these actions, the design may well require a major rethink.

In Figure 2.9 the memory cell has six transistors. This cell, often abbreviated as a 6T memory cell, has two back-to-back inverters and two access NMOS transistors. Since static RAMs comprise a large part of the area of a chip, great effort is generally taken to ensure that a compact 6T memory cell is available. CMOS manufacturers generally provide layouts for the 6T memory cells where compactness is achieved by allowing some layers to be closer to each other than is generally acceptable. These waivers in design rules ensure a small, economical memory footprint. Since the provided memory cells are very small, they have very limited drive abilities. To improve the speed of a memory it is possible for a designer to create a larger memory cell. This will, however, lead to a larger memory, especially if the CMOS manufacturer does not accept design rule waivers for this cell (determining which design rule waivers are acceptable requires running many process lots, an expensive and time-consuming process). The larger size will lead to greater parasitic loading, which will tend to counteract the effect of the greater drive strength. For these reasons most SRAMs, even high-speed ones, tend to use process-verified memory cells.

The 6T memory cell shown drives differential outputs. This is not true for all memories. Register files often have single-ended outputs. Since a register file's dimensions tend to be much smaller than a SRAM's and their speed requirements much higher, they often use much larger memory cells. Figure 2.10 shows the layout plot of a register file. The physical regularity of memories can be noted in Figure 2.10. This regularity often allows all the routing needs for the memories to be achieved by simply abutting appropriately designed cells. For memory designs, the primary design constraint is often efficient layout. An interesting observation in memories is that a single-ended register file is generally much faster than a larger SRAM with differential output sensing. This seems contradictory, since the push–pull action of a differential design should be faster than a single-ended approach. While this may be true in general, it has to be remembered that a large, heavily loaded differential SRAM has much more heavily loaded bit lines than a small, single-ended register file.

At the bottom of Figure 2.9 the sense amplifiers (often abbreviated to sense amps) are shown. In the figure the sense amps are shown as back-to-back inverters, which is how they can be conceptualized and indeed sometimes implemented. To improve speed, a current mirror can be used to provide extra current for faster switching. The sense amp is the only pure analog circuit that will be encountered in this book. It has to sense difference in voltage much lower than Vdd, often down to several tens of millivolts. This is made more difficult by the need to ensure correct operations across changes in process, voltage, and temperature. A sense amplifier is a differential amplifier in that it amplifies the voltage difference between its two inputs. While degradations for all transistors in the sense amplifier due to temperature and process variations are obviously not desired, they can generally be tolerated. Systemic differences between the two sides of the sense amplifier, however, tend to be more critical. Great care is taken in circuit design and layout to ensure that the two halves of the sense amplifier match each other. Physical design techniques include ensuring that the layout of the two sense amplifier halves is as symmetrical as possible, with all transistor having the same XY orientation. This ensures that any variations in the lithographic process used to manufacture the chip will impact the two halves of the sense amplifier similarly. In addition, sense amplifiers may use non-minimal gate lengths for sensitive transistors. While this will reduce the performance of the transistor, it does ensure that variations in drive strength tend to alter the absolute difference between the matched transistors less.

If some form of current source is used in a sense amp, care must be taken to ensure that this current is turned on only when sufficient voltage difference exists between the two inputs to the sense amp. This not only reduces power, but also ensures that erroneous voltage values are not sensed. A self-timed circuit, whose delay mimics the longest delay through the memory array, is customarily used to ensure that the sense amp is turned on when it should be. In addition to the constraints with matching transistors, sense amps are also constrained by the need to make sure that they fit together with the cells in the memory array. This requires each sense amplifier width to be an integer multiple of the width of the memory cell. For a register file with a full swing output signal, the sense amp is not an analog circuit and can be implemented with a CMOS buffer or some other purely digital circuit.

Like sense amplifiers, decoder cells and input drivers in a memory are constrained by layout needs. A decoder cell is generally only as high as a memory array cell. Memory decoders can use static or dynamic logic. In some applications it is necessary to access more than one memory location per cycle. This can be supported by allowing multiple read ports and write ports for the memory. This requires larger memory cells, more bit lines, and extra memory decoders.

From an ASIC designer point of view the primary tradeoff in memory designs is between speed versus power and area in different memory architectures, or speed versus memory size in different sized cuts of the same memory architecture. As in all other aspects of VLSI design, design-specific optimizations are often possible. Since this book is focused on the use of domino logic, the reader is reminded that if a memory provides data to a domino block then the only acceptable output transition for the data is from low to high during the evaluate cycle. This can be ensured by AND gating the memory output with the clock. This logic can also be incorporated in the sense amp or by using a domino-compatible flip-flop.

2.4 What to remember if applying domino logic

At the end of this chapter it is worthwhile to remember that as the designer prepares to apply domino logic to help with a seemingly intractable critical delay, some points must be verified. Firstly, the designer must be certain that standard static synthesis cannot achieve the desired results with a superior microarchitecture, or that such an approach does not lead to an unacceptable penalty in area or power. Secondly, if the critical path under consideration traverses through a memory, it should be verified that speeding up the logic will sufficiently help performance. There may be the need to use a different memory cut or change microarchitectural partitioning. Finally, care must be taken to ensure that the module under consideration has a reasonably "clean" interface. The inputs should preferably come from registers, especially if they are timing-critical. These flip-flops can then easily be transferred to domino logic-compatible ones. The output should preferably go to a flip-flop or a logic stage with sufficient timing slack to tolerate the extra latch inserted at the output of the domino cell. The module, and its timing challenges, should be well understood. Throwing faster logic at improperly understood logic is like throwing money at an improperly understood problem. Other than the fact that the money will be spent, few other outcomes are certain.

References

1. P. P. Gelsinger *et al.*, Microprocessors circa 2000, *IEEE Spectrum*, October 1989.
2. R. H. Dennard *et al.*, Ion implanted MOSFET's with very short channel lengths, IEEE International Electron Devices Meeting, 1973.
3. D. Harris, *Skew-Tolerant Circuit Design*, Morgan Kaufmann Publishers, San Francisco, CA, 2001.
4. M. S. Hrishikesh *et al.*, The optimal depth per pipeline stage is 6 to 8 FO4 inverter delays, 29th Annual International Symposium on Computer Architecture, 2002.
5. D. Chinnery and K. Keutzer, *Closing the Gap Between ASIC and Custom: Tools and Techniques for High Performance ASIC Design*, Kluwer Academic Publishers, Norwell, MA, 2002.
6. J. L. Hennessy and D. A. Paterson, *Computer Architecture: A Quantitative Approach*, Morgan Kaufmann Publishers, San Francisco, CA, Second Edition, 1996.

7. P. M. Kogge and H. S. Stone, A parallel algorithm for the efficient solution of a general class of recurrence equations, *IEEE Transactions on Computers* **22**(8), August 1973.
8. R. Hossain and L. B. Huang, System and method for predictive comparator following addition, US Patent Number 6820109 B2, November 2004.
9. R. Hossain, J. Herbert, J. F. Gouger and R. Bechade, A 5.2 ns cycle time floating point unit macrocell, 24th European Solid-State Circuits Conference, The Hague, Netherlands, 1998.
10. J. C. Herbert. R. Hossain and R. A. Bechade, Floating point unit having a unified adder–shifter design, US Patent Number 6148315, April 1998.
11. D. Markovic *et al.*, Methods for true energy performance optimization, *IEEE Journal of Solid State Circuits* **39**(8), August 2004.
12. M. Johnson, *Superscalar Microprocessor Design*, Prentice-Hall, Englewood Cliffs, NJ, 1991.

3 Domino logic library design

Razak Hossain and Thomas Zounes

3.1 High-speed digital circuit design

We start our discussions on designing a domino logic library by reviewing the answer to two classical results on sizing static CMOS inverters. While static and domino logic are different circuit families, they are both CMOS digital design styles, with the insight provided by studying static inverters being useful in understanding the general needs required for any library. The first issue relates to how the transistor sizes in inverters should scale to achieve a fast delay through a series of inverters driving a large capacitor. For example, if the first inverter has PMOS and NMOS transistor widths of 2 and 1 μm, what should the transistor sizes be in the next inverter? It seems obvious that the next inverter should have larger transistor sizes to ensure that the final inverter is strong enough to quickly drive the large load. The question that arises is how the transistor sizes should scale from one inverter to the next to minimize total delay. If the next inverter's transistor size increases quickly, it will heavily load down the inverter driving it. This will lead to a large delay. If, on the other hand, there is only a small increase in size between adjacent inverters then a very large number of cells are needed. Again, this will cause a large delay. The inverter sizing question leads us to think how different drives need to be sized. It also provides insight into the performance impact that occurs when we limit the number of drive strengths available, an inevitable consequence of using a standard cell library. The second question we will investigate is what ratio of PMOS to NMOS transistor widths should be used to minimize the delay through a set of inverters. Transistor sizing is the most basic step in circuit design, with proper sizing of different transistors being essential for good design and efficient layout. Taken together these two questions lead us to investigate, albeit only for static inverters, what drive strengths should be used and how each cell should be sized. This activity is essential whenever a standard cell library needs to be designed. Let us jump right in.

In 1975, Lin and Linholm compared the area versus speed tradeoff in sizing a set of cells used to drive a large capacitive load [1]. The intuitively obvious solution to the problem involves a set of progressively larger driver cells, so that the final cell is capable of driving the load circuit with a reasonable transition time. In [1] it is shown that by having a constant propagation delay across the different sized driver cells, the overall delay is minimized. Thus, while each cell has greater drive strength than the preceding one, it is also driving a larger capacitive load. In VLSI design large loads tend to be seen for nodes that have a high number of fan-outs or which are driving very long wires. Very

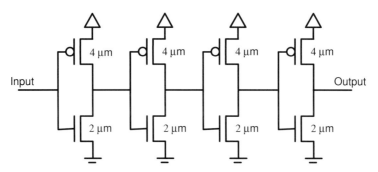

Figure 3.1. Four inverters in series.

large loads (for current CMOS processes something in the range of a few picofarads) requiring more driver cells are generally only encountered in the clock and reset tree and when on-chip signals have to be driven off-chip.

Richard Jaeger was one of the reviewers for [1] and used the authors' formulation for total delay to derive the result that the optimal transistor scaling factor for driver cells in the buffer system is e, or approximately 2.3 [2]. What this means is that if the size of the inverter transistor gate widths increases by 2.3 from each inverter to the next, then the total delay for the set of drivers is minimized. Jaeger also showed that delay variations are small around the optimal point. This slow roll-off of total delay from its optimal value is reassuring, since it means that by only using a limited number of drive strengths, as is typical in a standard cell library, it is possible to get a good approximation of the optimal result. In the years that followed, the driver sizing problem has been extended to include power, including short circuit power [3], reliability [4], and the effect of wire induction in CMOS processes [5].

The need to continuously increase the drive strength for cells in a logic path occurs when the final driver must drive a very long wire or many fan-out cells. If there is no need to drive a large capacitive load, increasing the drive strength of the cells by using progressively larger cells is not necessarily a good idea. Using larger drive strengths reduces delay, but the benefits of this are somewhat negated by the greater capacitive load encountered at the output of the cell due to the fact that larger cells will be present in the next stage. In addition, it is impossible to increase sizes indefinitely, or even to exclusively use maximum-sized library cells without leading to a very large and power-hungry design. Under such circumstances the delay minimization problem can be studied more fruitfully by focusing on the ratio of PMOS to NMOS transistor width. In Figure 3.1 four inverters are placed in series. Each inverter has a PMOS width, W_p, of 4 μm and an NMOS width, W_n, of 2 μm. The capacitive load seen by each driver is assumed to be the total gate capacitance of the next inverter's PMOS and NMOS gate widths, or $W_p + W_n$. If the transistors in the inverters are in the saturation region when on, the resistances of the PMOS and NMOS transistors are proportional to $1/(W_p\mu_p)$ and $1/(W_n\mu_n)$, where μ_p and μ_n are the hole and electron mobilities, respectively. Assuming that the mobility of electrons is twice that of holes, the delay for each inverter in Figure 3.1 is proportional to $1/4\mu_p$. This is true if the input is either rising or falling, since the transistors have been sized to compensate for the different electron and hole mobilities.

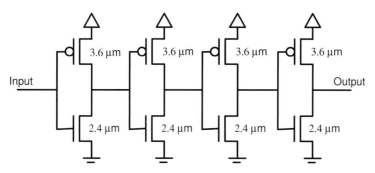

Figure 3.2. Four inverters with resized transistors.

As a next step let us change the sizes of W_p and W_n to 3.6 μm and 2.4 μm. This is shown in Figure 3.2. Since the total transistor width is constant, the total loading at the input of each inverter is the same as before. There has, however, been a shift in the relative delay for a rising and a falling transition. Since W_p is smaller, the rising delay has increased while the falling delay has been reduced. Since a falling transition is always followed by a rising one, the average delay through two inverters is reduced by about 3% compared with the original delay. The reason for this is that there is a greater reduction in falling delay than the proportional increase in the rising delay. The optimum ratio of W_p to W_n can be derived by differentiating the expression and is proportional to the square root of the ratio of electron to hole mobilities. In this example, where the ratios of the mobilities are assumed to be 2, the corresponding ratio of W_p to W_n transistor lengths should be 1.4. While this analysis uses a very simplified model of transistor delays (we also exclude interconnect and parasitic components of the delay), it does illustrate the principle that rather than striving for equal rise and fall delays, lower delay occurs by allocating greater transistor width to the faster mobility devices. Analysis for other cell topologies shows that the optimum PMOS to NMOS transistor width tends to vary, for example the ratio is different for an inverter compared to a three-input NAND gate [6]. Also, near optimum delays are possible even if the ratio varies within 5% of the minimum value. This provides some leeway, which is useful when laying out the cell.

Some observations can be made from the two topics that we have looked into. Firstly, in order to ensure good drive strength a gradual increase in the drive strength is needed. This ensures that wherever needed, proportionally larger drive strengths are available. The second point is that rather than striving for equal rise and fall times, it is better to emphasize the transition of the inherently faster transistor. This point helps to explain the advantage of domino logic. All logic transitions in domino cells occur only in one direction, with the domino cell internal node falling, which causes the output of the cell to rise. Most of the transistor width can hence be focused to emphasize this transition. Since the slow precharge delay does not propagate from cell to cell, it can be sized far less aggressively.

The discussions on circuit design have focused till now on transistor sizing. The other facet of digital circuit design at the cell level involves the use of different circuit topologies. For CMOS combinational logic implemented with dual PMOS and NMOS functions, there is little flexibility available since the implementation follows directly

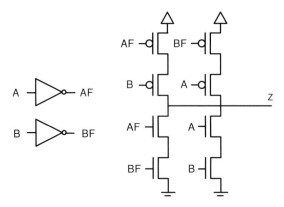

Figure 3.3. A two-input static logic XOR cell.

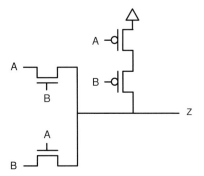

Figure 3.4. A two-input XNOR implemented with pass transistor logic.

from the boolean definition of the function. Often the only schematic choice is if an explicit buffer should be used on a large drive strength cell or if the designer should go with larger transistor sizes and no buffer. One design choice that may be available for a designer is to use pass transistor logic. In Figure 3.3 a static two-input XOR cell is shown. This design uses 12 transistors if one includes the two inverters needed.

In Figure 3.4 an NMOS pass transistor logic implementation of a two-input XNOR cell is given. Readers unfamiliar with this implementation should quickly check its validity. This design requires only four transistors. There are a number of other circuits which can be implemented similarly in a very compact way using pass transistor logic. In deeply scaled CMOS processes, however, complementary pass transistor logic has to be used to ensure that a good zero and one are propagated through the cell. For the two-input XNOR shown in Figure 3.4, this means that two extra PMOS transistors have to be added. In addition, two extra inverters are now needed to provide inverted signals. This quickly negates many of the advantages of a pass transistor implementation. There are two other disadvantages with pass transistor logic that limits their application in standard cell libraries. Firstly, pass transistor logic tends to be more area-inefficient to layout than standard CMOS stacks. The drain and source of a pass transistor generally need to be connected independently, requiring extra via connections. For standard NMOS and PMOS stacked transistors this is not the case. Secondly, if a cell input is directly connected

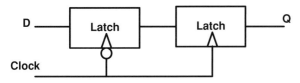

Figure 3.5. A master–slave flip-flop.

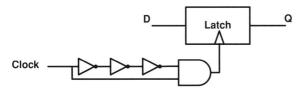

Figure 3.6. A pulse flip-flop built with a single latch.

to the source or drain of a pass transistor, then depending on the condition of the gate, the load capacitance seen by the input varies. This complicates cell characterization. It is for this reason that pass transistor logic is rarely used in standard cell libraries. When pass transistor logic is used in a standard cell, the input signals generally must be first buffered or inverted before being used.

While straightforward topologies are typically used to implement combinational logic, more funky implementations of flip-flops are increasingly becoming common. In fact, the largest change that we have seen in standard cell libraries over the last few years has been the increasing use of so-called glitch or pulse flip-flops [7, 8]. Before we describe these cells, let us quickly review a more traditional flip-flop design. The standard approach to designing edge-triggered flip-flops has been as master–slave latches. For a rising edge-triggered flip-flop this means that data enters the master latch when the clock is low. When the clock rises, new data becomes available at the output of the flip-flop. This scheme is shown in Figure 3.5. The total input-to-output delay through the cell (called the D-to-Q delay since the input of a flip-flop is generally called the D pin and the output Q) involves two latch delays.

A pulse flip-flop manages to reduce the D-to-Q delay for the flip-flop from two latch delays to a single latch delay. This is done by generating a small clock pulse on the rising or falling edge of the clock. The pulse turns the latch transparent, allowing data to be propagated through it. This structure, shown in Figure 3.6, allows the delay through the flip-flop to be reduced from two latch delays to a single latch delay. As pipelining has been used increasingly to increase the operating speed of designs, the delays through flip-flops have become a progressively larger part of the total cycle time [7]. Pulse flip-flops have correspondingly become very popular.

The single latch used for pulse flip-flops can either be a static or a domino implementation, with care taken to ensure that the precharge does not propagate outside the cell in domino implementations [8]. Care must also be taken to ensure that the generated clock pulse closely tracks the latch delay to ensure that both circuits are closely synchronized across changes in process, voltage, and temperature. The biggest danger in using pulse flip-flops is that they have longer hold times than standard flip-flops. This follows since the output of the cell can change throughout the width of the clock pulse. Extra delay buffers need to be used with these flip-flops to ensure hold time failures do not occur. Hold problems become more acute if the clock pulse driving the cells has a very long rise time, since this can adversely affect the quality of the generated pulse. While this may seem to be a relatively easy problem to avoid, in practice it requires verifying that every single pulse flip-flop in a design with perhaps tens of millions of instances, implemented by a number of different design teams, has a sufficiently fast clock transition. Like so much else in life, for VLSI design the devil lies in the detail.

3.2 An introduction to standard cells

Standard cell libraries provide a set of cells which are designed to achieve a good mixture of high speed, small area, and low power across a wide variety of designs. Custom design generally emphasizes performance for a specific design. Standard cells are designed to provide acceptable performance with high productivity. The constraints placed on standard cells are on their performance, their layout, and their need to interact with automated design tools (correct functional models and timing characterization).

A layout plot of a standard cell is shown in Figure 3.7. A power and ground line on the metal 1 layer (the lowest metal layer) are part of a continuous power stripe that provides energy to the cells. In between the power and ground line, the PMOS and NMOS transistors and connecting wires are placed. Each standard cell has a number of metal 1 routing layers, called tracks, available along its horizontal axis. These tracks allow for the node connections within the standard cell to be completed. Unused cell tracks are available for the router to use. The total number of tracks available in a standard cell library defines the largest PMOS and NMOS transistor layouts possible without splitting the transistor into a number of parts (the act of breaking up a large transistor into a number of fingers is referred to as folding the transistor. In computer architecture instruction, folding refers to merging a number of simpler instructions into a more complex one. The term, hence, is used to describe both splintering and merging actions). Having more tracks in a library helps standard cell designers by providing not only the possibility of larger transistor sizes, but also an easier way to connect different points in a cell. Unfortunately, this is not often possible since more tracks may lead to a larger design, which may not be competitive. Typically, standard cell libraries have between 8 and 15 metal tracks with the lower number of tracks being made available in less speed-critical, high-density libraries. Some other points to note about standard cells:

- Since standard cells are placed with alternating power (Vdd) and ground lines, every alternate vertical row of cells needs to be flipped to ensure that the cells are correctly connected to power and ground.
- Designing very large standard cells can lead to a cell in which it is difficult to connect all the nodes together. In addition to the metal 1, the other horizontal metal layers

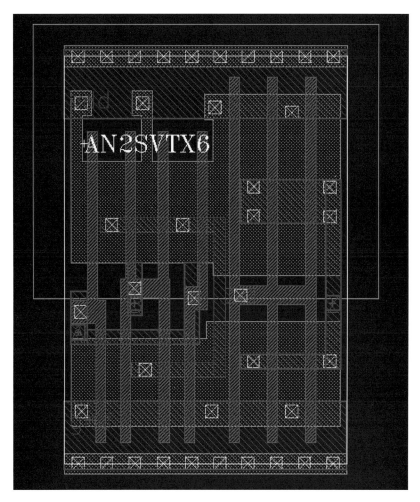

Figure 3.7. A standard cell layout.

(metal 3, metal 5, etc.) can sometimes be used to route connections within a cell. These layers are, however, used sparingly to ensure that enough routing resources are available for the automatic routing tool. In addition, large standard cells become very wide, making it difficult to place them efficiently. In most standard cell libraries, the largest cells tend to be registers or flip-flops.
- Most automatic place and route tools allow double-height standard cell layouts. These are standard cells with twice the height.
- Combinational standard cells cannot have multiple outputs. This is a disadvantage in domino logic, where many logic structures can be implemented efficiently as multiple output functions. The limitation stems from the inability of current synthesis tools to logically map such cells.

Many of the physical constraints on standard cells are based on the fact that they must be placed and routed with EDA tools. Standard cells are also synthesized with EDA tools.

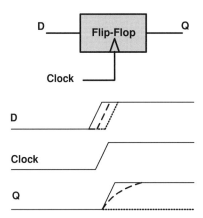

Figure 3.8. D-to-Q delay changes with smaller setup delays.

This tool-dependent use of the cells means that they must be optimized for the expected context in which they are used. Standard cell library development, hence, remains an iterative process involving circuit design, synthesis experiments, and place and route experiments.

A final challenge in developing standard cells is that each cell is an independently modeled instance. While a cell having an independent functional model seems reasonable, for timing analysis things get more complicated. Let us try to explain why. The timing analysis of ASIC-style design involves traversing through the design from inputs to outputs to determine the paths with the longest delays. For combinational logic cells this means that each cell must be characterized to ensure that every possible transition from every input to the output is measured. For a two-input AND gate this would mean that a path must exist from the inputs, let us call them A and B, rising, to the output, Z, rising. This must be measured under different output loads and input transition times. In order to ensure that Z will rise when input A rises, input B must be one. The difficulty in timing characterization is that a relationship exists between when the output rises and when input B becomes one. If input B rises a few picoseconds before input A rises then the output delay will be much longer than if B switches much earlier. Since the other input can change any time before the critical input arrives, current timing analyzers force a static value to be used to model an inherently dynamic situation. Generally, for characterization the other inputs are assumed to arrive much before the input being measured. In this case the characterized delay is optimistic. An alternate scenario may occur where parallel transistors in a standard cell switch simultaneously, leading to a smaller delay time than under characterization conditions where only a single input is assumed to change.

Sequential logic for which setup and hold times must be characterized have their own challenges. The definition of setup and hold time as mentioned in Chapter 1 is relative to a clock signal. It might be assumed that setup and hold times can be measured by repeatedly changing input or clock arrival times until the sequential element has a functional failure. This is shown in Figure 3.8, where an input is changed repeatedly closer to a clock rising edge. While failure occurs when a logic one is erroneously sensed as a zero and vice versa, this failure mode is not approached suddenly. Rather, the failure is initially manifest with correct transitions becoming slower and slower until they finally fail. Defining a setup or hold failure with a logic failure would mean that some extremely slow-changing outputs are possible, leading to likely timing problems in the next stage. For this reason, setup and hold times are generally not measured by a logical failure. Classifying a failure point based on the percentage degradation of the delay through the flip-flop is a better choice. While using the most realistic characterization scheme is always the best choice, if a new library is to be used with an existing library which has been characterized in a particular manner, one has to make sure that the two libraries use similar approaches. If this is not done, the delays in the libraries will be incorrectly calculated, leading to improper cell choices during synthesis and physical design.

3.3 Designing a high-performance standard cell library

Many of the optimizations required to design a high-performance standard cell library are part of the easy to understand, if difficult to define, category of good circuit design: having optimized schematics and layouts; an accurate characterization of timing and power; and ensuring that the cells are stable across variations in process, voltage, and temperature. The library performance will need to be validated against other available libraries. Two important activities encountered in the design of the library are defining the number and type of standard cells and choosing the drive strengths.

3.3.1 Starting the design

The most basic choice in designing a library, defining the library cell height in metal tracks, is often set by the competitive landscape in which the library must compete. The designer may, hence, have no flexibility in determining this. The one qualification attached to this situation is for the designer to be aware of what cells in an existing library are typically used. If designing with an existing standard cell library is pushing the limits of speed for a particular technology, the synthesis tool will be consistently using large cell drives. Under these circumstances, designing a new library with more routing tracks may allow for a more efficient implementation of the cells, without the need for extra buffering. Alternately, if the design is very difficult to design, the router will need a low utilization rate to finish its job. Here again it may be possible for a new library to be competitive despite being taller.

The next choice to be made is the number of cells that the library is to have. This is not often a hard number, since new cells and drives can be added throughout the development process, with the consequences of using the new cells being evaluated continuously. Typical numbers we have seen for general purpose libraries are around 500 standard cells. Smaller, specialized libraries focused on datapath cells, for example, may have only a dozen cells or so. A large number of the cells in the library are inverters and buffers, provided with many different drive strengths. The other cells tend to have far fewer drive strengths available. The number of cells in the library can mushroom quickly once the different permutations of flip-flops are considered: with and without scan; having dedicated or shared scan output; support for set, reset, or set and reset inputs controlled synchronously or asynchronously; inverted, non-inverted, or dual outputs. Thankfully, many of these variations can be included by changing a small part of the flip-flop without having to redesign the entire cell from scratch.

Academic papers have reported that increasing the number of standard cells in a library beyond more than a few dozen does not generally improve the speed of the design [9]. The author's own experience in designing libraries has shown that there are few speed advantages to a larger library after the first 50 or so most commonly used standard cells have been introduced. As more cells are added the delay tends to reduce, although even that is not guaranteed since adding a particular cell in the context of a specific design and synthesis tool may actually slow down the design slightly! This seems to be related to how the technology mapping process proceeds with different libraries. The major advantage with larger libraries is in area reductions that are possible. Anyone who has run a synthesis job knows that the most time-consuming part of the process is during the slow incremental optimizations, when more efficient local substitutions are attempted. Having a rich library helps this process. Excessively large standard cell libraries tend to slow down synthesis due to the very long synthesis runs encountered. In addition, design and characterization time for a library increases in proportion to its size. In recent years characterization has become more difficult due to the increase in the number of transistor models and environmental corners needed in new CMOS processes. Since the Spice models for a process tend to change often during the first few years of its operation, the characterization process has to be frequently rerun, requiring engineering resources to be assigned long after the design is complete.

3.3.2 Choosing drive sizes

Drive sizes for a high-performance standard cell library should be chosen rationally to allow the synthesis tool to efficiently implement the design. In this subsection we describe the approach used by us in the design of high-performance standard cell libraries. The approach focuses on the maximum expected load that each cell will drive.

The delay of any standard cell is proportional to the output load that the standard cell is driving. In addition to the load-dependent component of the delay, there is an intrinsic delay associated with the cell. This delay is due to the fact that a certain delay is required to traverse through a cell even if the output load is zero. Increasing the size of a cell leads to greater intrinsic delay – since the transistors of the cell are larger, they have larger input capacitive loads (a point to remember during circuit design is to ensure that the inputs of the cells being designed are driven by other cells and not the Spice simulator directly. A Spice-like simulator will have an infinite drive capacity, which will mask the delay increase due to using larger input transistor sizes). In Figure 3.9 the delay through cells with drive size of $1\times, 2\times, 3\times$, and $4\times$ is shown as a function of output loading. For

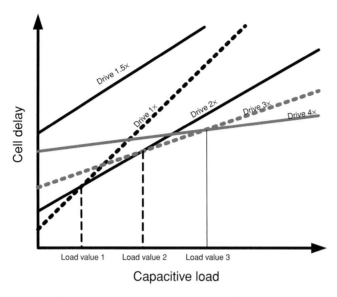

Figure 3.9. Drive strength versus delay for a set of standard cells.

the 1× cell the delay is minimum when the output load is below "Load value 1". After this the 2× sized cell is fastest up to "Load value 2". Drive 3× is faster from here till a capacitive load of "Load value 3", after which drive 4× is fastest. This illustrates that there is here no intrinsically fast cell and slow cell, rather the drive sizes reflect speed as a function of output loading. The drive 1× size is determined by layout and the likely load encountered by a cell driving a single fan-out placed close to the cell. The next few loads should reflect typical loading conditions encountered by cells along a critical path. Knowledge of the parasitic and fan-out loading conditions will determine good values for the other load values at which a transition should be made to a larger load value. This is process technology-dependent. Typical values for these increments seen in a 90 nm process have been at 30 fF, 60 fF, and 90 fF. For very large loads it is assumed that the output of the cell will be buffered, so there is no advantage in using very large drive strengths. Drives 1×, 2×, 3×, and 4× are assumed to be sized for use on the critical path of the design, a few other drive strengths can be developed for non-critical paths. Such a drive is shown as drive 1.5× in Figure 3.9. The cell is not the fastest cell for any loading condition, but it can drive much larger loads faster than a drive 1×. Having such a cell will mean that a larger, more power-hungry cell does not need to be used when a non-critical path is heavily loaded.

The values of the different cut-off points (Load value 1, Load value 2, etc.) should be fairly consistent across the cells of a library. This simplifies the delay optimization process for the synthesis tool. A very low input capacitance load (0.5×) is possible for those cells that are frequently part of large fan-out cones. The designation of 1×, 2×, 3×, etc. for the drive strength of the cells is a somewhat arbitrary moniker, a 2× cell transistor may not be exactly twice the size of those of a 1× cell. To ensure that the output load of a cell does not greatly violate the load value it was designed for, the standard

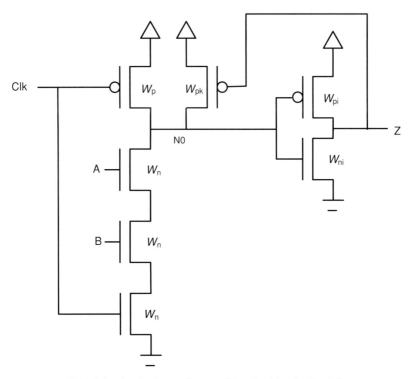

Figure 3.10. CMOS domino logic two-input AND cell with a feedback keeper.

cells can be characterized with a maximum output load. This is particularly important when domino logic cells are considered.

The authors were involved in the design of a high-performance static cell library in a 0.18 µm process. Let us recount some of our experiences and design choices. After some experimentation, the effective PMOS to NMOS ratio chosen was around 1.5 [10]. The actual ratio varied slightly for each cell depending on layout compactness. We discovered that larger non-buffered cells led to better performance than using smaller cells and buffers. Most cells had four drive strengths, with the more popular cells other than inverters and buffers having up to eight more sizes. Inverters and buffers were provided with even more drive strengths. The library achieved a 20% speedup, validated on silicon, over an existing high-speed library, which met our requirements and was very satisfying!

3.4 Circuit design of domino logic cells: a qualitative approach

Domino logic cells are optimized for fast rising output transitions. In Figure 3.10 a two-input domino AND gate, first shown in Chapter 1, is reproduced. We have already discussed how in static logic the sizes of the PMOS and NMOS transistors are intricately linked, with increases in the relative size of PMOS transistors negatively impacting fall times, and vice versa. In domino logic, since the logical transition is always rising, the sizes of the NMOS pull-down transistor, W_n, and the PMOS pull-up transistor, W_{pi}, can be made large for fast transitions. The speed of precharge in the domino design depends on the complementary transistor sizes, W_p, the pull-up transistor size, and the inverter NMOS transistor, W_{ni}. During precharge W_p charges the internal node, causing the output to fall.

When designing a domino logic cell one of the first constraints that needs to be considered is the maximum operating speed of the logic in which it will be used. This determines the time available for precharge. For a 1 GHz operating frequency with a clock duty cycle of 50%, 500 ps is available for precharge. From this, variation in clock duty cycle and clock skew must be subtracted. Setting a very low precharge delay, let us say, to assume a maximum operating frequency of 2 GHz for the logic with worst-case process, voltage, and temperature will mean that a significant portion of the total available transistor width is assigned to precharge delays. Unfortunately, this will reduce the evaluate phase delay of the cell, leading to a slower design. If the actual maximum operating frequency of the design is much lower, the speed penalty in using an overly aggressive precharge delay may lead it to not being sufficiently fast compared with a static logic library.

The other explicit transistor seen in Figure 3.10 is the weak feedback keeper transistor W_{pk}. This transistor is sized to be weak, often by setting its transistor length to be larger than a minimum-sized value. The purpose of the weak feedback transistor is to keep the output of the cell at 0 when the cell does not switch. Without a weak feedback the output state of the transistor would be kept by virtue of the load capacitance on the internal node of the transistor. This capacitance can be altered by leakage current through the transistors. Furthermore, the capacitance on the load can be diminished by charge sharing and crosstalk noise. We discuss these topics next.

3.4.1 Charge sharing

Charge sharing refers to the reduction of charge on the internal node of a cell under certain input switching conditions. The condition is illustrated in Figure 3.11, where a three-input domino AND gate is shown. During the evaluate phase let us consider the case where inputs A and B turn high while C remains low. If the nodes Na and Nb had originally been zero, then turning on A and B leads to some of the charge on N0 being distributed across these two nodes. The final voltage on the node N0 depends on the ratio of the capacitance of nodes N0 to Na and Nb. Charge sharing can lead to the output node of a cell being so diminished that it flips the output state of the cell. In Figure 3.11, if all input went high, obviously node N0 would go low; this is different from charge sharing, where the node is discharged unintentionally.

Charge sharing is countered by the weak feedback in the domino logic cell. This transistor provides charge that will cause the voltage level of the internal nodes N0, Na, and Nb to rise. There is a tradeoff between how charge sharing-tolerant a domino logic cell is (which requires a stronger weak feedback and a weaker pull-down NMOS stack) versus the rising delay of the cell (which prefers a weaker feedback and larger W_p transistors).

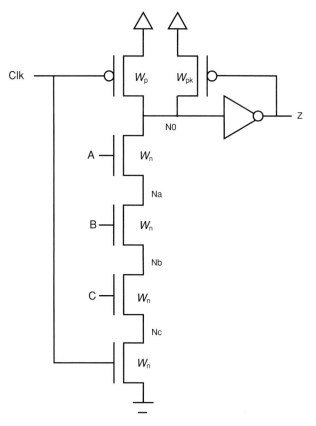

Figure 3.11. A three-input domino logic AND gate.

This tradeoff can be avoided in custom design frameworks, where the evaluate signal arrives in a narrowly defined window of time, by turning off or weakening the feedback transistor as the evaluate phase starts. Since we are designing domino logic to be used in a more automated framework, such timing relationships cannot be assumed, especially at low clock frequencies. Another approach that has been proposed to counter the effects of charge sharing is the use of extra integral precharge transistors. We encountered two problems with the use of such transistors in the design of domino logic circuits. Firstly, the layout of standard cells was greatly complicated by the inclusion of internal pull-up transistors. Secondly, since the charge-sharing input might not switch until late in the evaluate cycle, we noticed that much of the internally stored charge was lost when low threshold voltage transistors were used. This loss of current through leakage is strongly dependent on the process corner and temperature used for simulations.

3.4.2 Crosstalk noise

The classical definition of noise in electronics refers to the deviation of devices from their idealized models due to the non-continuous, quantized nature of charge. Shot noise and thermal scattering are standard examples of this. In digital design, noise is used as an umbrella term to refer to a collection of real-world complexities that creep into designs, with crosstalk often being the biggest concern.

Crosstalk occurs when energy from a signal on one line is transferred to a neighboring line by electromagnetic means. Crosstalk is usually defined in terms of aggressors and victims, where the aggressor transfers some of its energy to the victim. In general, both capacitive and inductive coupling exist. On-chip, however, the currents through the signal lines are usually too small to induce significant magnetic coupling. While inductive effects, such as simultaneous switching noise, may grow in importance in future process generations, they are still manageable with careful routing of the power and ground network. Capacitive coupling, on the other hand, depends on the line-to-line spacing between wires and can be measured as the total capacitance of a node that is coupled to neighboring wires, as opposed to fixed sources. For domino circuits, crosstalk can lead to high pulses being induced on lines that should be low. These pulses can cause the domino cells to switch erroneously, leading to possible wrong states in the design.

As CMOS processes scale, wires have become progressively narrower. To limit the resistance in them, they have also become taller. This has meant that most of the capacitance in wires is coupled to adjacent wires on the same level. Aggressor and victim pairs are, hence, most likely to occur on long wires routed next to each other. The actual occurrence of crosstalk is very complicated, being dependent on the strength and location of the aggressor and the victim lines (crosstalk is worst when aggressor and victim lines are driven from opposite sides), the ratio of floating to fixed line capacitance (greater floating capacitance causes crosstalk-induced bumps to be higher), the resistance of the line (higher resistive lines weaken aggressors), the degree to which the victim may already be weakened due to other aggressors (if an attacker simultaneously attacks the line driving the victim, the strength of the victim driver is reduced, making it more difficult to overcome crosstalk noise on its output), and the location and number of aggressors present (multiple aggressors can attack a line, for example, if a bus is routed together). Crosstalk can be manifest as a high and narrow pulse on the victim line, or as a wider but lower signal. While crosstalk analysis can become very complicated, for ASIC domino logic design it can be reduced to two problems. The first is determining the expected crosstalk-induced pulse on the cell inputs. The second is characterizing domino cells to determine the maximum crosstalk pulse that the cell can tolerate on any input. Current place and route tools can estimate the maximum crosstalk noise bump on the inputs of cells. If every cell input is characterized by a maximum bump that can be tolerated, the tool will determine possible crosstalk failures and attempt to reroute the problem wires to achieve a violation-free implementation. To ensure that no noise violations occur in cells, their crosstalk susceptibility needs to be characterized conservatively. As EDA tools continue to improve their crosstalk analysis and avoidance capabilities, less conservative modeling techniques can be used.

3.5 Circuit design of domino logic cells: a quantitative approach

The design process for a standard cell library is an iterative process in which, as cells are added to the library, concurrent synthesis experiments on benchmark circuits are used to evaluate the benefits of the new cells. For domino logic design, the comparisons will include a comparison with the results achieved using a static library. The design process therefore entails a set of delay measurements for every cell as it is being designed. This data is used for a direct comparison and also to provide timing characterized views of the cell that can be used in synthesis. In this section we describe in detail how we performed timing characterization for a domino logic library designed with a 90 nm CMOS process. Domino logic cannot have inverting outputs. This means that while NANDs, NORs, and inverters cannot be used, ANDs, ORs, and AND–OR functions can be. Since it is possible to have a wide variety of AND–OR functions, the number of types of possible cell is very large. In Chapter 4 we describe how inverting logic is avoided when domino logic cells are synthesized.

Before plunging into a description of the characterization step, let us define some terminology that will be used. Data inputs refer to all the inputs of the cell except for the clock, while the term "all inputs" includes the clock. The term "pin under test" (PUT) refers to the specific pin which is being tested. Related pins refer to the pins in addition to the pin under test that need to be high to ensure that the output can switch. For example, in a two-input AND gate, if input A is the pin under test, then the related pin B must be high to ensure that the output of the gate can go high.

The design of a domino logic library involved ten characterization tests for all non-register-type domino cells. These are:

- Cell delay and output transition time measurement.
- Input pin capacitance measurement.
- Setup measurement of data input rising relative to the clock falling.
- Minimum pulse width high overlap (MPWHO) measurements.
- Hold measurements of data inputs falling relative to the clock rising.
- Setup measurements of data inputs falling relative to the clock rising.
- Minimum clock pulse width for low and high phases.
- Maximum noise spike characterization for the input pins.
- A charge-sharing check.
- Precharge sizing check.

3.5.1 Cell delay and output transition time measurement

The delay from the inputs switching to the output switching determines the performance of the cell. This delay depends on the input signal rise/fall time and the load on the cell. For domino logic cells the output can fall only when the clock goes low and the cell can rise only if the clock is high and the data input logic evaluates high. For this reason the rising delay is measured for all inputs, while delay falling is measured only when the clock falls. Since our domino logic flow assumes that domino cells can interact with static cells, the trip value to indicate where the measurement begins and ends is the same as those used in the available static standard cell library. This reduces errors when mixing

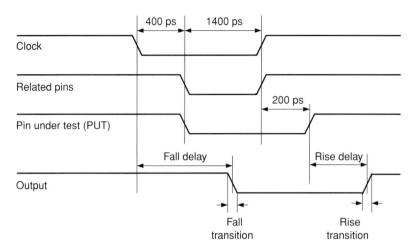

Figure 3.12. Delay and transition time measurements for a domino logic cell.

domino and static cells. For a falling signal the trip value is 60% of Vdd (as a start or end point). When a signal is rising, the trip point is 40% of Vdd. Transition times are measured from 20% to 80% of Vdd.

Figure 3.12 shows the waveforms used to measure delay for a 90 nm domino logic cell. Initially, all inputs are kept high. This causes all internal nodes of the cell to be low. The clock is then forced low, and after 400 ps, all the data inputs are also driven low. Between the time the clock falls and the input pins fall, the internal nodes precharge. This 400 ps precharge time allows the internal nodes to be precharged to what we consider a "reasonable" value. Let us explain the purpose for this. In a domino logic design, the internal or evaluate node that drives the inverter is high during precharge. The data inputs eventually go low during precharge (this depends on the phase of the driving cell), since the driving cells also enter precharge. There is, however, a lag between when the clock enters precharge and when the data inputs go low. During this time, internal nodes start to precharge to an NMOS transistor threshold voltage below Vdd (Vdd – Vtn). This extra charge on the internal nodes increases the delay in a domino circuit. The amount of time that the input nodes are on will determine the charge on the internal nodes. In some cases, the internal nodes may remain at 0 due to the input signals remaining low. In other cases, the internal node may precharge to Vdd – Vtn. A precharge time of 400 ps is used during characterization since the library has precharge delays in the range of 300 ps and 400 ps. We believe this value reflects a realistic condition without being excessively pessimistic or optimistic. Precharge delays vary with the process corner, so a lower value would be more realistic when a non-worst process, voltage, and temperature (PVT) corner is used.

The output of a domino only goes low when the clock falls. The fall delay for the cell is measured from the falling edge of the clock input.

Output rising delay measurements are made for every cell input, including the clock. In many cells more than one combination of inputs can cause the cell to evaluate. In such cases the input condition that leads to the longest delay is selected. To measure the rise delay, the related pins rise with the clock. All other data pins remain off. The PUT rises 200 ps later. This 200 ps ensures that all the pins of the cell do not switch at the same time, and represents an attempt to provide a typical condition in which the PUT will rise. For rise delay measurement from the clock pin, all relative pins rise at the same time as the clock.

The delay and transition simulations are done for five different output loads and five different input transition times. Synthesis and timing analysis tools can interpolate all the needed values based on these data points. The maximum input and clock transition used for characterization is 150 ps (a 150 ps transition from 20% to 80% of Vdd is equivalent to a 250 ps 0 V to Vdd transition). The maximum cell load used is the maximum capacitance value for the particular drive. This maximum capacitance is the capacitance load that leads to a 400 ps (0 to 100% of Vdd) falling transition for a domino cell. The falling transition is used since this is greater than the rise transition in domino cells. Since the maximum capacitance value depends only on the output inverter size, it is constant for all cells of the same drive strength and does not need to be characterized independently for every cell.

3.5.2 Input pin capacitance measurement

The input pin capacitance is measured by taking the integral of the current of the PUT's driving source and dividing it by Vdd. This follows as the charge stored on a capacitor is equal to its capacitance multiplied by the voltage across it. Current is measured only when the signal is rising, and is recorded for each of the 25 delay measurement runs. The average of all these values is then used.

3.5.3 Setup measurement of data input rising relative to the clock falling

It is possible for an evaluate phase signal being propagated through a set of domino gates to become progressively thinner until it finally leads to a logical failure. This failure is manifest as a pulse that is too narrow to allow the domino cell to evaluate. The pulse nature of domino signals means that checks must be performed to ensure correct pulse width. One mechanism by which the output of a domino cell can have an excessively thin pulse is if an input signal rises just before the clock falls. In order to ensure that the output pulse is appropriately wide, a setup time is defined between the data inputs rising and the clock falling.

Since the purpose of this setup time is to ensure proper pulse propagation, the setup is defined to occur when the output fall delay decreases by 2.5% compared with the fall delays generated by a very large data setup time. The decrease in the precharge delay is due to the internal node of the domino cell not falling all the way to 0 V during the evaluate phase. While this definition may appear to be very conservative, it is needed to ensure output transition times do not vary much from that measured during delay

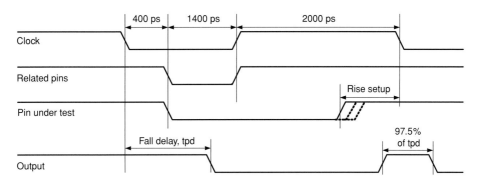

Figure 3.13. Setup rising versus clock falling measurement for a domino cell input.

characterization. If the data setup time is allowed to become smaller the output rise transition times will increase, with the output possibly failing to reach Vdd. Ensuring that output fall delays are within 2.5% of the possible maximum delay ensures a limited impact on the transition times of the domino cells.

The simulation approach used to calculate the setup rising delay is similar to that for the rising delay measurement as seen in Figure 3.13. The clock initially goes low and related pins fall soon after. Related pins and clock then rise simultaneously, and after a delay, the pin under test rises just before the clock falls. This delay time when the PUT rises is then varied until the output fall delay is 97.5% of the original fall delay. This defines the PUT rising setup.

Like delays, setup and hold times are dependent on the transition times of the nets that drive the inputs and the output load of the cell. The rising setup time is characterized by varying clock fall times, data input rising times, and output loading. A total of nine simulations are run for every input data pin. The timing tool interpolates other values from this data.

One of the advantages of controlling pulse shapes by a setup check is that this attribute is understood by ASIC tools, ensuring that synthesis and physical design tools can correct setup failures during the implementation process. Unfortunately, it is not possible for all checks required by domino cells to be understood by the tools, as we shall see with minimum overlap checks described next.

3.5.4 Minimum pulse width high overlap characterization (MPWHO)

When multiple inputs are applied to a domino cell care must be taken to ensure that the pulses overlap each other sufficiently to allow the evaluation node to be discharged. For example, in a domino AND3 gate all three inputs must be on at the same time for the internal node of the cell to be discharged and the output of the cell to rise. The minimum pulse width high overlap (MPWHO) is a check to ensure that the overlaps between the different data inputs to a domino cell are acceptable. This is needed for all inputs to the domino cells. In a domino OR3 cell the transistors are connected in parallel. Here the

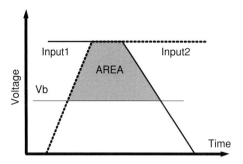

Figure 3.14. Minimum pulse width high overlap area.

check merely needs to determine that each input applied to the cell is sufficiently wide to cause the output of the cell to rise. In some AND–OR cells, many different paths can cause the output of the cell to rise. Since we consider each domino cell to have only a single MPWHO value, the worst-case overlap value is used.

For our 90 nm domino logic library, the MPWHO is defined as the duration of time between when the last input rises (measured at 40% of Vdd) and the earliest input falls (at 60% of Vdd) for each NMOS pull-down path such that the evaluation node falls to 20 mV or lower. From simulations the 20 mV evaluation node voltage is found to ensure that the output delay does not increase more than 1% under maximum capacitance loading.

Since pulse width overlaps are not standard timing checks, they are not supported natively by Synopsys's PrimeTime^{TM} or other timing analysis tools. This required a definition of MPWHO to be developed for the domino logic library. Defining the minimum overlap needed between inputs based on measured time proved difficult, since many inputs can have slow transition times. Forcing the synthesis tool to try and fix all transition times to achieve acceptable high-value overlap for all cells would unnecessarily overconstrain the design. In order to avoid this, a metric was developed to model overlap between different inputs under very different transition times. This metric measured MPWHO based on the total input signal overlap area for each cell, above a cell-specific fixed voltage value, Vb. While this choice may appear somewhat arbitrary, this solution married a simple characterization process with excellent correlation of the evaluation node reaching 20 mV or lower.

Figure 3.14 shows how the MPWHO area for a domino cell can be measured. For a two-input domino cell, let us assume that Input2 is the last rising input and Input1 is the first falling input. The intersection of these signals above Vb gives the minimum pulse width high area overlap. The voltage Vb ensures that unless the input signals cross a certain value, the output will never change. Since NMOS transistors only turn on when their inputs exceed a threshold voltage, this is a reasonable assumption. The MPWHO constraint can be checked for every domino cell in a design with Synopsys's PrimeTime^{TM} and some Tcl scripting. The script reads a file with the MPWHO area and Vb data for every input. Delay and transition time information available in the

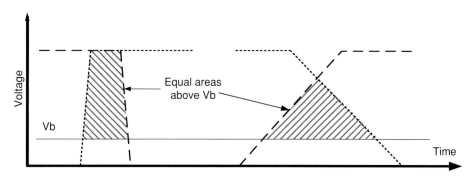

Figure 3.15. Setting Vb to ensure equal MPWHO area under slow and fast transition times.

timing analyzer can then be used to determine if a violation occurs. Needless to say, this approach requires that all domino cells in the library have a characterized MPWHO area and Vb value.

In order to ease characterization of the MPWHO area for each domino input, simulations were limited to two cases: under very small and very large transition times. This was deemed acceptable as experiments showed that the worst-case MPWHO conditions occur either when the nets of a cell are very lightly or very heavily loaded. If the pins are driven by lightly loaded nets, fast rising and falling signals can lead to minimum signal overlap. Alternately, if the input pins are heavily loaded then slow rising transitions can limit the area over the Vb voltage level. Under worst-case process and environmental conditions, minimum transitions are simulated with a 12 ps transition time, while the maximum transition uses a 150 ps transition time. For each set of Eldo^{TM} runs, the rising input is moved closer to the falling input to reduce the area overlap as shown. Once the evaluation node fails to read 20 mv, the minimum overlap condition is assumed to have been reached. The data from the fast transition and slow transition times are then used to find the Vb that will cause the areas to be the same. This is shown in Figure 3.15, where Vb is set such that the two simulations have identical MPWHO areas. From Figure 3.15 it can be seen that the cell inputs do not all need to be simultaneously at Vdd for the output of the cell to switch.

For an OR gate the rising signal and falling signal is applied to the same pin. For gates that have more than one transistor in the evaluation path, the transistor connected to the evaluation node will have its gate connected to the falling signal and the adjacent transistor will have its gate connected to the rising signal. This is shown in Figure 3.16. All other transistors in between will be connected to Vdd. If there are other transistors not part of the evaluation path, their gates will be connected in such a way as to cause the longest evaluation time of the evaluation path. The above condition ensures worst-case characterization. This ordering of inputs and associated logic transitions limits the characterization effort needed for MPWHO.

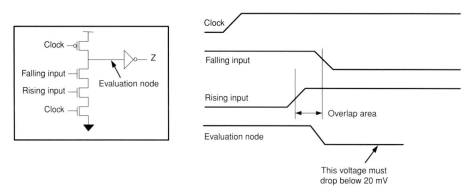

Figure 3.16. MPWHO measurement setup for a domino cell.

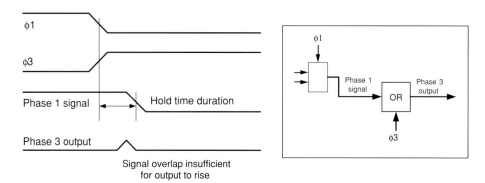

Figure 3.17. Hold violation for data fall with respect to the clock rising.

3.5.5 Data pin hold falling measurement

In a synthesized domino design some cells may have inputs driven by cells connected to a different clock phase. Under such circumstances a cell input may go low when the cell clock rises. This is shown in Figure 3.17. While the clocking scheme tries to avoid this circumstance, it may still occur, and hence, all cells need to be verified against this failure mechanism. This failure mechanism is in fact a classical hold time failure and is avoided with a data fall to clock rise hold check for all inputs. This check can be considered to be an overlap check, but the clock pin is not considered in the MPWHO checks (for which the clock input is kept high) since it was found that considering an extra serial transistor in MPWHO checks made the check overtly pessimistic.

The definition of hold time is the time between clock rising (measured at 40% of Vdd) and data input falling (measured at 60% of Vdd) that will cause the evaluation node of the PUT's cell just to drop below 20 mV. The related pins are chosen to cause the longest delay for PUT (as in the delay characterization). The hold check is done using two extreme clock rise times and two extreme data fall times (four data points). Output loading is set to the maximum cell loading to ensure worst-case tolerance. The measurement used for data pin hold with respect to clock falling is shown in Figure 3.18.

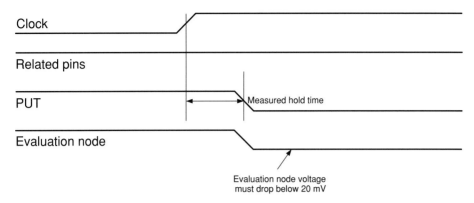

Figure 3.18. Domino cell hold time measurement.

3.5.6 Data pin setup falling measurement

A requirement in domino logic is that the output of all cells must go low every cycle before the clock rises again. If this does not happen, a false evaluation may occur. To detect such an issue, a setup check is performed to make sure that the data input goes low on each cell before its clock rise. The definition of this setup is the time between data input going low and clock going high, such that the evaluation node is charged high.

Simulations determined that this check tends to always pass. For this reason a computationally simple procedure is used to measure the setup time. The setup time is defined by measuring the distance between the input and clock pin of the cell so that they intersect at 20% of Vdd. This is a conservative number that ensures the data input transistor and the clock transistor are both off as the data input falls and the clock rises. This simplified check is suitable for the test cases encountered by us in our 90 nm designs. For other processes or designs a more thorough characterization process may be needed. One of the reasons that we always wish to use simple characterization wherever possible is that it allows for the library to be characterized faster. While using parallel simulations does allow the characterization time for a library to be speeded up, the high cost of EDA licenses and computer servers limits the number of parallel simulations that can be run.

Two input data pin transition values and two values of clock rise transition times (four points) are used to build the setup table. Figure 3.19 graphically illustrates the setup time measured.

3.5.7 Minimum clock pulse width for low and high phases

It is possible for a domino cell to meet all the setup and hold times but still fail due to the clock duty cycle being severely distorted. This is manifest as a very narrow high or low value for the clock. For clocks provided directly from a PLL, the probability of this occurring is low. Still, as a verification check this needs to be tested, especially since clock duty cycles can vary considerably if they are generated by dividing a faster clock source. Since this check does not require great precision, it is implemented

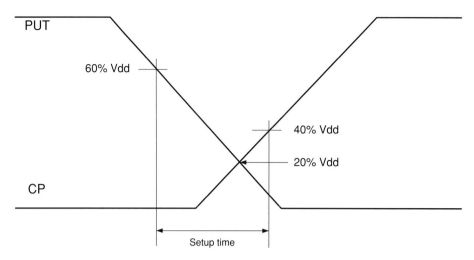

Figure 3.19. Data fall to clock rise setup time measurement.

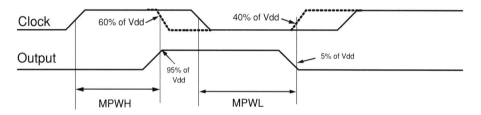

Figure 3.20. Clock minimum pulse width characterization.

using the existing minimum pulse width checks supported by the Liberty format timing models (often called the Synopsys.lib model). The primary limitation with using the minimum pulse width high (MPWH) and minimum pulse width low (MPWL) check is that it provides a single number for an input pin. In reality, the minimum clock pulses that are acceptable depend on the clock transition times and the output loading on the cell. For our purposes this check is characterized assuming the maximum clock transitions specified in the library and the maximum output loading for the cell in question.

Figure 3.20 shows the MPW characterization timing. The minimum pulse width checks are performed by initially setting high the data pins that cause the longest evaluation delay. The clock is then forced high. The MPWH is measured from when the clock reaches 40% of Vdd, until when the output of the domino cell reaches 95% of Vdd. MPWL is measured from when the falling clock reaches 60% of Vdd until when the output reaches 5% of Vdd. Simulations showed that the output reaching 95% of Vdd led to almost the same delay as the clock falling to 60% of Vdd. Using the output rise point to measure delay allowed us to avoid an iterative process of measuring acceptable clock pulse width by sliding the clock window and replacing it with a single pass measurement.

3.5.8 Data pin maximum noise spike characterization

A major concern with domino logic is its susceptibility to crosstalk-induced noise failure. Such noise can lead to functional failures if the evaluation node in the domino cell is inadvertently discharged. To ensure noise tolerance of domino cells, every input of every cell is characterized for its crosstalk tolerance. This characterized data is then used during physical design.

For the 90 nm domino library, the chosen place and route methodology involved using Synopsys's Astro™ physical design tool. In Astro™ it is possible to specify the maximum acceptable noise spike on each input pin of every cell. The router will then attempt to ensure that this noise spike constraint is not violated. A limitation with the version of the tool we were using was that the noise spike was calculated based only on direct aggressors. This did not allow for the tool to consider the effect of noise being propagated from other cells. The tool also did not enforce any constraints on the width of the noise pulse. Since the total crosstalk-induced energy is proportional to both the height and width of the induced voltage bump, it should be considered for crosstalk tolerance of the design. To overcome these limitations the domino cell characterization process assumes worst-case pulse width and that propagated noise is also present. In addition to crosstalk, simultaneously occurring charge sharing in the domino can cause the internal node to dip. As the evaluation node dips the output inverter's NMOS transistor becomes weaker, and hence less able to counter induced crosstalk noise on its line. For this reason charge sharing is assumed to occur simultaneously with crosstalk. Characterizing the cells under such pessimistic conditions ensures the functionality of the cells. It also goes some way to explaining why silicon results for digital ASIC designs tend to run faster than the simulations!

The maximum spike characterization involves applying up to two input spikes to the cell inputs. One of these spikes is assumed to be on the PUT, while the other is to another input pin. The second input spike is added to include the possibility of multiple crosstalk events being simultaneously applied to the domino cell. The maximum charge-sharing condition is applied simultaneously with the spike noise. The input noise and charge sharing weaken the domino cell output inverter NMOS transistor. On this weakened output, capacitively coupled noise is applied to the output node. The same output noise is applied independently to the same cell with no input noise. The output voltage area above 10 mV is calculated for the cell with and without noise. If the ratio of these two values is more than 3, then the input noise is too large and a smaller applied bump is applied until the ratio of 3 is reached.

The value of 3 is used for measuring acceptable noise bump area since up to that value, if the noise bump is applied to another domino cell it is not seen to propagate to the output of that domino cell. The domino cell is thus able to filter the input bump. One of the possible problems with noise is that it cannot be allowed to propagate from cell to cell since it could ultimately overwhelm a cell further along the logical path. The value of 3 is very design- and technology-specific. For our purposes we discovered that using values greater than 3 led to possible noise stability issues, while smaller values appeared to be unnecessarily conservative. The characterization procedure is shown in Figure 3.21.

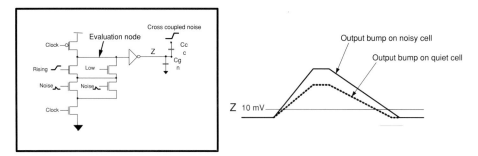

Figure 3.21. Maximum voltage spike simulation setup.

The 90 nm domino library has a maximum capacitance setting on all cell outputs that limits the worst-case transition time to 150 ps. For simulation, the aggressor cross-coupling ratio is assumed to be 50%. Post place and route experiments showed that 85% of nets have less than 50% of total capacitance coupled to all other nets combined. These two factors, the maximum transition time of signals and the maximum cross-coupled capacitance connected to an aggressor, limit the width and height of the worst aggressor noise that can be applied to the output of the domino cell being characterized. For an aggressor having a rise time of 300 ps, the input noise bump will be very low, but the width will be at its maximum. When the aggressor rise time decreases, the bump will also increase, but the width will decrease. Different aggressor rise times are generated by having an inverter drive a variable capacitance. The capacitor value is varied until the output area ratio is 3. Simulations have shown that worst-case crosstalk tends to occur with low, but wide, aggressor pulses.

To simulate worst-case charging during crosstalk measurements, the simulation is set-up to have all internal nodes discharge. Great effort was spent ensuring that the worst-case charge-sharing condition and input noise was applied during cell characterization (it was not always obvious which input pins would cause worst-case noise). The maximum noise spike is also measured for the domino cell clock pin. The clock network has a large capacitive load being driven by a strong set of drivers, so only minimal noise spikes are expected on it.

3.5.9 Charge-sharing check

In addition to the characterization for noise sensitivity on the input pins, a number of checks are included to ensure the effectiveness and stability of the domino logic cells. One of these is the charge-sharing check. One should note that while charge sharing is assumed during the worst-case voltage spike simulation, the worst-case charge-sharing scenario may not be checked since some inputs need to remain low to measure crosstalk. To check the impact of charge sharing, the maximum charge-sharing condition is applied to the cell, and its stability is determined. If the cell fails, the cell is redesigned to ensure stability. As mentioned earlier, changing the size of the weak feedback keeper allows for the cell to tolerate charge sharing better. Making the weak feedback stronger is not often acceptable, since it will increase the delay of the cell. If a domino cell is not sufficiently faster than the equivalent static cell (domino cells should in general be 1.5× faster than the equivalent static cell), the cell is removed from the domino logic library. Other techniques to reduce the charge-sharing effect of the cell include having a larger output inverter (this increases the evaluation node capacitance), increasing the ratio of the PMOS to NMOS transistors in the evaluation stack (effectively increasing the capacitance of the evaluation node with respect to the other internal nodes), and changing the topology of the cell. Topology changes involve changing the relative location of transistors in serial stacks and are most useful when multiple pull-down stacks of transistors are present for large-sized drives. By switching the relative order of input in different parallel pull-down stacks, it is possible to increase the charge-sharing tolerance of the cell without altering its functionality.

3.5.10 Precharge sizing check

The precharge PMOS transistor in domino cells must be able to precharge the node effectively to ensure correct operation. The precharge sizing check involves making sure that the internal node of the domino cells reaches at least 90% of Vdd at maximum operating frequency. Possible problems can be corrected by increasing the size of the precharge PMOS transistor.

3.6 Characterizing domino logic-compatible registers

As mentioned earlier, domino logic designs do not require the use of explicit flip-flops. Scan testing is, however, mandatory for many digital designs. This, along with the desire to introduce hard clock edges in domino synthesis, means that a domino logic flip-flop can be useful.

Domino flip-flops differ from static flip-flops in that resets are synchronous and that the output of the flip-flop can change both for the evaluation phase, when the clock rises, and for precharge, when the clock falls. Static flip-flops only change during a single clock edge. To support scan testing, the register may also have a dedicated scan out (SO) output. This output is similar to the output of a static logic register and maintains its output value for the full clock cycle.

The domino registers are designed as pulse flip-flops. As mentioned, pulse flip-flops have shorter setup times, but significantly longer hold delays. The input of the domino register is a single latch stage, with the output having an inverter and a feedback keeper like a standard domino logic cell. The latch is transparent for a short period of time after the clock rise. A bare bones domino logic-compatible flip-flop, without scan, is shown in Figure 3.22. In this design, if the evaluation node goes low, it remains low

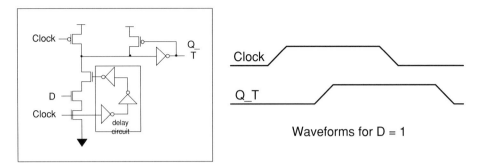

Figure 3.22. A basic domino logic-compatible register.

as the NMOS transistor cell leakage is greater than that for the PMOS transistor. For processes where this is not true, extra transistors can be used to ensure the stability of the design.

Like other domino cells, the domino register requires an input pin capacitance and a maximum output load characterization. In addition, three additional characterization tests must be run for a domino-compatible flip-flop:

- Clock to output delay and transition characterization.
- Data input setup characterization.
- Data input hold characterization.

3.6.1 Clock to output delay and transition characterization

In domino logic, flip-flop delay is measured from the clock rising to the output rising, and from the clock falling to the output falling. This differs from a static flip-flop in which both output rise and fall delay is measured from the rising clock edge. The scan output for the cell is a static output. For the scan output, the possible delay transitions are the same as a static flip-flop. As with other domino cells, the related pins are chosen to cause longest delay. During delay measurement the output transition times are also measured under different capacitive loads. All transitions assume a 150 ps worst-case rise or fall transition measured from 20% to 80% of Vdd.

The characterization approach is illustrated in Figure 3.23, where the measured delays for the domino and scan outputs are shown. The transition times of the outputs under different loading conditions are measured at the same time. The scan output delay is used directly in the characterization table. The delay values for the QT and QF output are, however, increased by 10%. This increase is used to compensate for the fact that setup times are measured assuming a 10% degradation in the output delay of the flip-flop (the setup measurement is detailed in the next subsection). Of course, if the domino signal inputs arrive early, the 10% delay is not incurred. Under those circumstances the domino registers have an extra delay penalty.

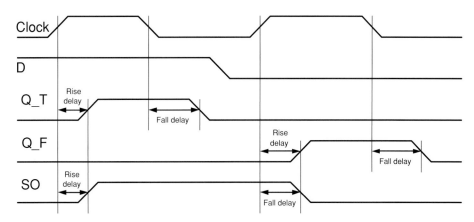

Figure 3.23. Domino logic register delay characterization.

3.6.2 Data input setup characterization

As with static registers, but unlike regular domino cells, the domino register requires data inputs to rise before the clock rises. The setup time for domino flip-flops is measured as the delay at which point the output pin delay is degraded by 10% compared with a signal that arrives much earlier than the clock. In addition to meeting the 10% delay increase, the minimum rise setup time must also ensure that the evaluation node reaches 20 mV (it is completely discharged when the output rises) and that the fall setup time does not dip below 90% of Vdd (when the output does not rise). The related pins are chosen for setup measurements to ensure the largest setup value.

3.6.3 Data input hold characterization

Input data needs to remain stable for a period after the clock goes high. The hold time for domino registers is generally much larger than that for non-pulse-based static registers. In Figure 3.24 the hold measurement scheme is shown for a register with true and false outputs. The hold simulation setup is very similar to the setup characterization. The only difference is that the related pins are chosen to produce the fastest delay. The criterion for failing is a 1% increase in output delay. A hold time failure is a serious issue that can lead to functional failures under any clock frequency or operating conditions. It is for this reason that a much more conservative measure for hold failure is used.

3.7 Layout of domino logic standard cells

For the 90 nm domino logic-compatible library the design process was strongly coupled with the efficient layout of the cells. For domino logic cells speed is achieved by maximizing the size of the NMOS transistors in the pull-down stacks. In order to facilitate this

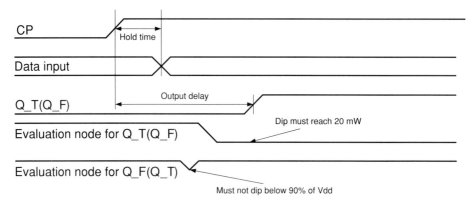

Figure 3.24. Domino register hold time characterization.

requirement, the n-wells within the cells are bent. Large NMOS transistors are placed on the left-hand side of the cell, along with the smaller PMOS keeper and precharge transistor. Large PMOS transistors are only needed at the output inverter of the cell. The output inverter is positioned on the right of the cell. The layout for a domino logic two-input AND gate is shown in Figure 3.25.

All domino logic cells are abutable with standard static cells. This is important since the domino logic flow uses both static and domino cells which may be placed next to each other. For large drives, it was often necessary to use multiple transistor fingers to achieve sufficient drive. When this occurred for cells with long pull-down chains, such as three-input and four-input AND gates, flipping the input sequence in duplicate stacks often led to less charge sharing and better noise tolerance. As mentioned earlier, the domino logic library did not use any intermediate precharge transistors. The layout was also greatly complicated by the presence of intermediate node precharge transistors.

3.8 Timing models for domino logic cells

Much of this chapter has dealt with the characterization process used to model a domino logic-compatible standard cell. In Figure 3.26 a complete timing model in the Liberty format is shown. The model is for a domino buffer. While all the complexities of modeling a domino cell are not shown in the figure, it does still give a flavor for how we deal with domino logic cells.

The timing model for the buffer begins with an instantiation of the cell. After the area of the cell is provided, the first domino-specific attribute is_dyn is defined. Domino logic-specific constructs are highlighted in italic in Figure 3.26. Since the domino logic library contains both domino and static cells, the attribute is_dyn is used to specify if a cell is a domino design. Traversing down the model we see that the function of the cell is defined to be the product of the clock and the input, A. The Liberty

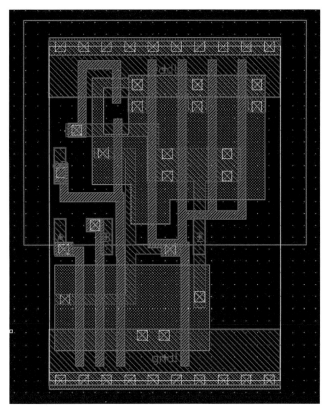

Figure 3.25. A domino logic standard cell layout.

timing model specifies the relationship between the output and the related input pins. For input A, the relationship with the output is specified as combinational rise, ensuring that the input cannot cause the output of the cell to fall. This reflects correctly the functionality of a domino cell. Delay and transition tables for pin A rising to the output are then provided (DOM_BFHST1F10R10_td). For brevity these values are not shown. The timing relationship between the clock pin (CP) and the output (Z) show rising and falling paths. The two input pins A and CP are seen to have an input capacitance and direction attribute defined. This is similar to a standard static cell. For pin A, setup constraints are also defined. Non-sequential setup times can be defined within the Liberty timing format. Setup times with respect to the clock falling and rising are provided for input A.

Timing models such as that shown in Figure 3.26 are generally understood by timing analyzers. Use of timing analyzers is, however, generally deferred until late in the design flow. During the implementation process, a simpler model must be provided to the synthesis tool. In Chapter 4 we will discuss these simpler models, along with an ASIC flow for using a library of domino logic cells to implement digital logic.

```
cell(DOM_BFHST1F10R10) {
  area : 32.1376 ;
  is_dyn: true;
  pin(Z) {
    direction : output ;
    function : "CP*A"
    max_capacitance : 0.2196 ; timing() {
    timing() {
      related_pin : "A" ;
      timing_type : combinational_rise;
      timing_sense : positive_unate ;
      cell_rise(DOM_BFHST1F10R10_td) {....}
      rise_transition(DOM_BFHST1F10R10_td) {....}
    }
    timing() {
      related_pin : "CP" ;
      timing_type : combinational;
      timing_sense : positive_unate;
      cell_rise(DOM_BFHST1F10R10_td) {.....}
      rise_transition(DOM_BFHST1F10R10_td) {....}
      cell_fall(DOM_BFHST1F10R10_td) {....}
      fall_transition(DOM_BFHST1F10R10_td) {.....}
    }/* end timing */
  } /* end pin Z */
  pin(A) {
    direction : input ;
    capacitance : 0.00568 ;
    timing() {
      related_pin : "CP" ;
      timing_type : "non_seq_setup_falling";
      sdf_edges : both_edges ;
      related_output_pin : Z ;
      rise_constraint(DOM_BFHST1F10R10_dsur) {
    ....}
    }
    timing() {
      related_pin : "CP" ;
      timing_type : "non_seq_setup_rising";
      sdf_edges : both_edges ;
      fall_constraint(DOM_BFHST1F10R10_dsuf) {
    ....}
    } /* end timing */
  } /* end pin A */
  pin(CP) {
    direction : input ;
    capacitance : 0.00810 ;
    min_pulse_width_low : 0.4221 ;
    min_pulse_width_high : 0.2200 ;
  } /* end pin CP */
} /* end cell DOM_BFHST1F10R10 */
```

Figure 3.26. Liberty format timing model for a domino buffer.

References

1. H. C. Lin and L. W. Linholm, An optimized output stage for MOS integrated circuits, *IEEE Journal of Solid-State Circuits* **SC-10**(2), April 1975.
2. R. C. Jaeger, Comments on 'An optimized output stage for MOS integrated circuits', *IEEE Journal of Solid-State Circuits* **SC-10**(3), June 1975.
3. H. J. M. Veendrick, Short-circuit dissipation of static CMOS circuitry and its impact on the design of buffer circuits, *IEEE Journal of Solid-State Circuits* **19**(4), August 1984.
4. B. S. Cherkauer and E. G. Friedman, A unified design methodology for CMOS tapered buffers, *IEEE Transactions on Very Large Scale Integration (VLSI) Systems* **VLSI-3**(1), March 1995.
5. Y. I. Ismail and E. G. Friedman, Effects of inductance on the propagation delay and repeater insertion in VLSI circuits, *IEEE Transactions on Very Large Scale Integration (VLSI) Systems* **8**(2), April 2000.
6. D. S. Kung and R. Puri, Optimal P/N width ratio selection for standard cell libraries, 1999 IEEE/ACM International Conference on Computer Aided Design, San Jose, CA, 1999.
7. D. Harris, *Skew-Tolerant Circuit Design*, Morgan Kaufmann Publishers, San Francisco, CA, 2001.
8. F. Klass *et al.*, A new family of semidynamic and dynamic flip-flops with embedded logic for high-performance processors, *IEEE Journal of Solid-State Circuits* **34**(5), May 1999.
9. M. N. Duc and T. Sakurai, Compact yet high-performance (CyHp) library for short time-to-market with new technologies, 2000 Conference on Asia South Pacific Design Automation, Yokohama, Japan, 2000.
10. N. Richardson *et al.*, The iCORE™ 520 MHz synthesizable CPU core, 39th Design Automation Conference, New Orleans, LO, 1998.

4 Domino logic synthesis

Razak Hossain and Bernard Bourgin

4.1 Introduction to domino logic synthesis

In the earlier chapters of this book we have seen that domino logic is intrinsically faster than static logic. The logic family is, however, more complex to use since every cell is clocked. Furthermore, the cell outputs are only valid during the evaluate phase, with the precharge phase resetting the cell. With domino logic the designer has to consider not only the logical functionality of the circuit, but also the clocking scheme. Domino logic design has traditionally only been available to those design groups who have an absolute need for high speed and can afford to utilize large numbers of engineers to handcraft circuits using this design style. This approach to domino logic design has meant that design productivity associated with the use of domino logic, measured in terms of cost and turnaround time (TAT, the time needed to complete a task) has lagged that of automated static logic. While the quality-of-results (QoR) generally improves with custom design, this may still lead to an unfavorable tradeoff in terms of cost versus benefit. For many design groups a fully automated solution provides adequate or close to adequate results.

The dynamic behavior of domino logic is part of the challenge in using it. At high speeds the clock and data are involved in a complex timing interplay which must be resolved correctly for proper functionality. The data for every domino cell must be propagated before the precharge signal arrives. In addition, the nature of the logic family makes it more susceptible to a number of different failure mechanisms including charge sharing, crosstalk, and power bounce, which are generally not problems in static logic. All these potential risks need to be monitored and formally checked. An automated domino logic design system is capable of checking all specified domino logic failure mechanisms, ensuring a highly reliable solution.

The two main advantages of automated design flows, improved engineering productivity and greater design reliability, remain valid when applied to the use of domino logic. These two factors improve the predictability of the implementation process, allowing for greater confidence when applied to the design of a large SOC. Project schedules for large ASICs specify, often many months ahead, the exact dates on which different blocks have to be available. With custom design this becomes difficult to do. Without an ability to provide predictability with the use of domino logic, it will be much less likely to be used, especially if the inability to complete the domino modules in a timely manner imperils the whole project.

Domino logic synthesis 71

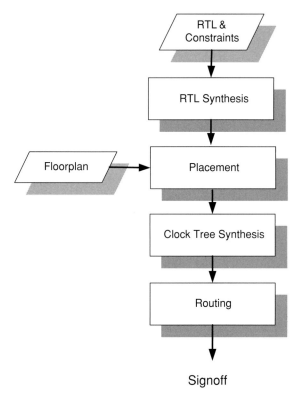

Figure 4.1. A typical static logic ASIC design flow.

4.1.1 A standard tool-based approach to domino logic synthesis

There are two broad approaches to domino synthesis which we categorize as ad hoc and standard tool-based. An ad hoc domino synthesis system can be constructed with custom tools and flows applied to a dedicated or standard hardware description language (HDL). This approach allows designers to achieve very good results, approaching that of a handcrafted solution. This approach is, however, costly since the tool user needs a good understanding of the domino logic design methodology. This approach also requires considerable EDA tool support to maintain and update the custom software. The second approach, which we will present in this chapter, relies on off-the-shelf tools, design flow, and existing HDL specifications of the design. A slight degradation of performance has to be borne by reusing the tools and design skills that are needed for static design. We do believe, however, that it is a more practical solution when using domino logic for most design groups.

Figure 4.1 shows a typical static logic ASIC implementation flow. The input to the flow is the synthesizable RTL description of the logic, generally in Verilog or VHDL, along with a set of timing constraints and synthesis directives, and some physical floorplan constraints. The main steps are: synthesis, which will transform the RTL into a standard cell netlist; placement, to place the netlist into the floorplan; clock tree synthesis, which

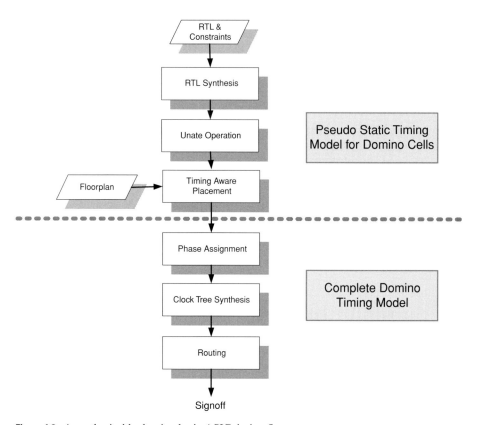

Figure 4.2. A synthesizable domino logic ASIC design flow.

will balance the clock pins; and routing, to physically connect all logically connected points.

Figure 4.2 shows how a static logic ASIC-style implementation flow can be extended to support domino logic synthesis. At first glance the domino synthesis flow has added two domino-specific steps: the unate transform and the phase assignment step. We shall describe these steps in more detail, following a quick description of how the other existing steps are modified when domino logic is used.

As mentioned, for domino logic synthesis we wish to reuse the exact same RTL, timing constraints, and synthesis scripts (these are generally Tcl-based synthesis commands) as used in static synthesis. Doing this involves mapping the RTL to the domino logic standard cell library. There are two reasons why standard synthesis tools do not support domino logic synthesis. Firstly, the tools do not understand the clocking scheme inherent in domino logic and its interaction with the logical functionality of the design. Secondly, they are not capable of mapping a standard cell library that does not contain inverters or other inverting cells. In order to overcome these difficulties and allow the tool to use the domino logic cells, the domino logic library is provided in a pseudo-static form. To do this, the clock pin is removed from combinational domino logic cells to keep only their boolean function. The precharge timing arcs (when the output falls) are replaced by the evaluate phase timing arcs (when the output rises). Negative unate functions are created by copying the positive unate equivalents, i.e., NAND cells are defined by copying the timing behavior of AND cells, NOR cells from OR cells, inverters from buffers, etc. Finally, dual rail binate functions (exclusive OR cells and multiplexers) are converted into single rail equivalent functions. The result is a library that has similar timing arcs and functionality to a traditional static library, only being faster and with identical rising and falling timing arcs. This library is used in the synthesis of the initial timing-driven placement step.

During the phase assignment step the pseudo-static models are replaced by the actual domino model. The clock pins are then connected to the clocks, assuming the use of a number of overlapping clock phases [1]. Clock tree synthesis has the same objective as with static logic, except that now in addition to the sequential elements the clocks must also connect the domino gates. This creates a massive increase in the number of clock pins to be balanced. Further complicating clock tree synthesis is the need to synchronize both edges of the clocks and also to maintain correct latencies between the different clock phases. We have been pleased to discover that commercial EDA tools are able to achieve acceptable results with this challenging task.

In routing a domino design the physical design tool faces the challenge in routing a very large clock tree network. Since domino logic tends to be more sensitive to noise than static logic, additional steps need to be taken during routing to minimize crosstalk. Current routing tools are capable of crosstalk prevention by wire spreading and performing a quick analysis to identify potential static noise violations (voltage bumps). For static logic, crosstalk violation correction is fairly simple since the voltage bump threshold tends to be the same for all signal nets, with threshold perhaps being higher for clock nets. For domino cells, on the other hand, all the inputs have been characterized by a maximum voltage bump which the routing tool must meet. In order to correctly optimize for both the evaluate and precharge phases, the routing tool must understand the full domino timing model. This model must also be understood by the signoff static timing analysis tool.

We shall now discuss in more detail the two unique steps needed in domino synthesis: the unate transform of the logic and the correct phase assignment for the domino cells.

4.2 Unate transform

Since domino logic cells cannot have inverters or inverting logic present, all such inversions need to be moved to the primary inputs or outputs of the domino logic module. For a logic block with multiple pipeline stages, the inverters can also be incorporated into flip-flops driving the logic. The actual movement of inverters through logic can be achieved by the recursive application of De Morgan's law [2]. Since inverters are schematically represented in logic as bubbles, the unating process by which inverters are moved to the inputs of the module is also called bubble pushing. In Figure 4.3 a circuit is shown before and after bubble pushing. The initial circuit is seen to have inverters between logic cells. After bubble pushing, all the inverters are placed at the inputs of the circuit. At

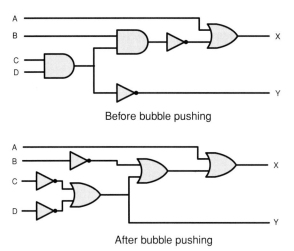

Figure 4.3. Removing an inverter with bubble pushing.

this point they can be easily incorporated into domino-compatible flip-flops driving the logic.

Unfortunately, it is not possible to always push inverters to the inputs of the block. Under those circumstances it may be possible to remove the inverters by pushing them forward to the outputs of the module [2]. This is not possible when an inverted and uninverted form of the signal is part of some reconvergent logic. When this occurs the inverter is considered to be trapped. To remove trapped inverters one must duplicate the logic cone from the trapped inverter to its primary inputs [2]. The overhead in extra cells needed to remove trapped inverters, with its corresponding increase in power and area, is the main disadvantage with logic duplication. In Chapter 1 it was mentioned that, providing extra timing slack exists, inverters can be tolerated by inserting a flip-flop and using a hard edge to introduce the inverted and non-inverted copy of the signal at the appropriate time. To use this technique requires very tight control of clock delays and skews. This is difficult to do in an ASIC framework unless the design has very long cycle times.

The example in Figure 4.4 shows the unating process for a design containing AND cells, OR cells, and inverters. Indeed, for domino synthesis it is possible to construct functionally correct designs by initially mapping the combinational logic to these cells and then applying the unating process. For production devices this approach is not, however, recommended as it has a number of limitations. Firstly, the synthesis of arbitrary functions to only these three logical gates leads to a sub-optimal implementation compared with directly mapping the design to a complete cell library. Secondly, the solution assumes that logic is combinational. Partitioning a complex design into combinational and sequential blocks is time-consuming and forces the domino logic to be synthesized differently from the static logic. The process can also lead to errors.

In order to transcend these problems and provide a more efficient solution, domino logic synthesis is allowed to directly utilize all the domino cells in the library. Since the unate process requires the use of De Morgan's law, the pseudo-static library is also

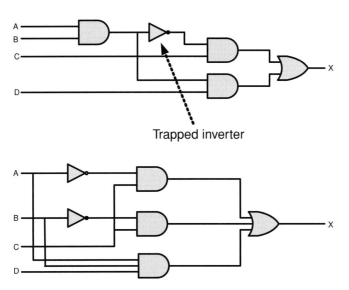

Figure 4.4. Removing a trapped inverter with logic duplication.

provided with the dual for every domino cell in the library. A domino cell without a dual cannot be included, since bubble pushing will not be able to proceed for that cell. In addition to specifying the dual of every library cell, additional attributes are added to the library to help with the unate transform. These attributes identify the nature of the cell pins (dynamic or static, inverting or not) in the library.

In addition to inverting cells, a dummy exclusive OR (XOR) and exclusive NOR (XNOR) are also provided during initial synthesis. During bubble pushing the binate XOR cell is substituted by an actual domino cell with a duplicated input fan-in cone. Initially providing the dummy model ensures that the domino logic uses XOR functions, a known difficult problem in domino synthesis [3]. The bubble-pushing algorithm continues the back-tracking algorithm until it encounters flip-flops. The outputs of flip-flops are treated as inputs for the next pipeline stage of logic, i.e., points from which the inverted and uninverted version of a signal is available.

While flip-flops are not needed for synchronization in domino logic, they are allowed for easy synthesis of existing register transfer level (RTL) code and to enable the design to be scan testable. As mentioned previously, flip-flops introduce "hard edges" [1] which limit performance somewhat. The penalty for using flip-flops in synthesizable domino logic is reduced by using domino flip-flops with very low setup delays. Since the flip-flops drive domino cells, they enter precharge when the clock falls. In order to maintain the stored data when the clock is low, they have a static slave latch. This latch is used for scan testing and ensures that the stored data can be retrieved even after the output of the flip-flop enters precharge.

4.3   Phase assignment

The second domino-specific step needed in domino synthesis is phase assignment. During phase assignment the pseudo-static library is replaced by one having the full domino timing model. The clock pin is now visible in the domino cells and needs to be attached to the proper clock phase. Since timing data is very important in the phase assignment procedure, this task is done after placement so that actual wiring delays can be considered. Attaching clock phases to domino cells not only specifies the timing relationship of the final design, but also determines if the design is functional. It is possible to apply a number of different clocking schemes to domino designs. For our purposes a four-phase clocking scheme (with each clock having a 50% duty cycle) is the preferred domino clocking methodology as it reduces the skew requirements on the clock [1].

What makes the phase assignment process difficult is that the synthesis process is performed only under worst-case evaluate cycle conditions. This is necessary as no current, commercially available synthesis tool can understand the dual operating mode of domino logic. Using the evaluate cycle delays to perform synthesis is a reasonable assumption, since it is the critical condition for well-conditioned domino circuits. Nevertheless, the precharge cycle is also important, especially when we start including non-clocked cells in the design. Complicating the matter further is that precharge delays must be considered for both the fast and slow process and environmental corners, to test for hold and precharge failures, respectively. Thus, a design must be phase-assigned to ensure that the final circuit works correctly in two different operating modes (evaluate and precharge) under different operating conditions (worst and best). During phase assignment, timing analysis is done to ensure that all the known failure mechanisms are detected. While a certain number of corrections, such as changing gates from static to domino logic or adding buffers are acceptable, a large number of repeated checks and alterations should be avoided. This is to avoid globally altering the initial synthesis and placement results. After phase assignment the design is routed, which will cause the parasitics values to change from the placed estimates. The phase-assignment process must be stable enough to tolerate these alterations without requiring extensive changes as there is no easy way to iterate across the entire design space. The solution to the phase-assignment problem, hence, must be such that it will lead to designs that are very likely to be stable moving forward, i.e., they are close to correct by construction.

This leads to the natural question as to what a correct domino design is. It is a design that should satisfy three conditions: firstly, it must be functional from a very low speed to the maximum operating frequency; secondly, it must be fast (specifically faster than an equivalent static design, otherwise why design in domino); and thirdly, it should be an implementation that consumes the least area, power, routing resources, and other standard metrics. It is somewhat meaningless to say that area is more important than power, or some other criteria, since requirements will differ across designs. Any measure to reduce the area or the power consumption of the design is welcome, provided it does not affect the possible functionality of the design. This is necessary as synthesized domino logic is a new technique, with any silicon failure likely to trigger a severe aversion to using the technique in the future. Also, due to the ubiquitous presence of the clock, certain failure mechanisms such as hold time failures, if tolerated in the phase-assignment script by using margins, are statistically more likely to lead to a problem than in static design. The three criteria, and their relative importance, should be remembered when sophisticated phase-assignment optimizations are considered. These optimizations generally involve trading off power or area for greater risk, and as such are not as important as guaranteed functionality and speed of operation.

In order to ensure that the domino logic design implemented with multiple clock phases is functionally correct, the following guidelines should be followed in the phase-assignment process:

- All logic paths in a domino logic design must proceed monotonically from phase $\phi 1$ through phase $\phi 4$. The clock phases are labeled $\phi 1$ to ϕn. They are all assumed to have equally spaced phase shifts. David Harris, in his book *Skew-Tolerant Circuit Design* [1], does not recommend allowing any phase skipping, where two logically connected domino cells have a phase difference of greater than 1. While we tend to avoid phase skipping in domino synthesis, it is not forbidden.
- No inverters or inverting logic can be used in logic paths, other than immediately after a static input or just before a flip-flop. This follows directly from the basis of domino logic. We do, however, allow inverters on signals driven by special static inputs, such as in the reset and clock tree networks.
- The static input signals must be clocked by a domino gate which rises after the static signal is stable, i.e., the input delay of the signal, plus the worst-case clock skew between the modules (or more strictly, when the static falling edge is stable). If a static input is stable at or after 75% of the clock cycle, it must be clocked at a flip-flop and not a domino cell. Static output signals must not receive a domino precharge signal that can be sampled. A static primary input or output signal is one that is received from, or sent to, a module implemented with standard static logic.
- The domino primary inputs are specified by a maximum rise delay and a minimum fall delay. This delay also includes the effect of intermodule skew. The constraints file specifies which clock phase drives the domino input signal.

If the specified rules are followed, a domino logic implementation will be functional. The design may not be able to operate at the maximum synthesized frequency, due to minimum overlap failures or precharge failures, but we are getting close to functional domino synthesis flow. If static logic cells are also present in the design, the design may have to be further slowed down to correct for problems such as precharge failures. Ensuring the design remains functional and can still be operated at the maximum clock rate follows if certain phase-assignment rules for domino logic are followed.

4.4 Phase-assignment rules

The phase-assignment process for domino cells assigns an initial preferred clock phase to each cell based on the arrival time of the last rising data input. Cells with inputs arriving in the first quarter of the clock period will be clocked by phase $\phi 1$, while if the inputs arrive in the second quarter of the clock period they will be clocked by phase $\phi 2$, and so on. Such a straightforward scheme has some problems which need to be corrected. We describe these problems next, along with how they can be overcome.

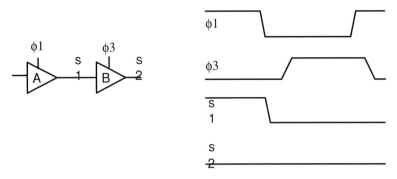

Figure 4.5. Illustrating a phase skipping-related error.

4.4.1 Phase skipping in domino cells

Domino cells on a non-critical, weakly driven path may have a driver cell on phase 1 and the receiver cell two phases later. Phase skipping of two phases involves a potential race between the clock skew of the two cells versus the precharge delay of the driving cell. Figure 4.5 illustrates how skew can lead to a high value at s1 not being captured in cell B. This is a hold time failure.

Risk evaluation. Since all the initial synthesis is done to meet worst-case process and environmental conditions, a quick risk evaluation can be made by asking if the problem described is more serious for typical silicon. If so, it is a serious problem. A similar, but slightly different, formulation of risk evaluation can be made by asking if the problem is less likely to occur when the design is run slower. If this is true, it is reassuring as the problem may be masked by slowing down the design.

With faster transistors, phase skipping is more likely to result in a functional failure. This is because faster silicon will decrease the precharge time of the cells (which makes the solution more risky). While clock skew is also reduced with faster transistors, this has a less direct and a smaller effect than the reduction in precharge time. Assume that the silicon is at the same process corner, but running the design slower is neutral for skipping two phases, as the problem is a hold time failure. When skipping more than two phases, running the clock slower is more likely to lead to a failure. This follows as the precharge value has more time to propagate and be sampled.

Problem detection. The hold falling constraint with respect to the rising clock detects this problem. Indeed, this is the reason for this characterization step described in Chapter 2.

Problem solution. During the phase-assignment process the sequence of cell for every path is checked. There are two solutions to this problem. Inserting a buffer (domino or static) may resolve this problem. A domino buffer should be clocked with the missing phase. Sometimes it is possible to avoid adding an extra buffer by readjusting the clock phases of the domino cells along the critical path. Thus, if two domino cells are on phase $\phi 1$, followed by a domino cell on phase $\phi 3$, it may be possible to clock one of the domino cells on phase $\phi 1$ by $\phi 2$. This will require timing slack to exist, which is generally present

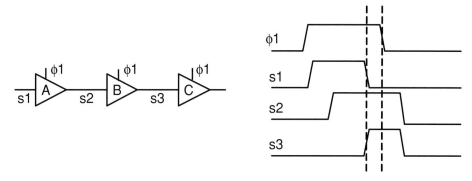

Figure 4.6. Data rising setup check with respect to the clock falling with a non-balanced clock phase assignment.

in this situation as it is a hold failure (hold problems are a manifestation of the next edge coming too quickly. Alternately one can think of them as existing when insufficient logic exists between adjacent registers).

In general, increasing the duty cycle of the clock helps with most evaluate cycle error mechanisms. For designs employing only domino cells increasing the duty cycle can be advantageous, since it helps solve evaluate cycle errors. Increasing the duty cycle, however, reduces the precharge time available. This can lead to other problems.

4.4.2 Unbalanced phase assignment

Unbalanced phase assignment can lead to an error in which the critical path of the design is no longer from a primary input or the output of a flip-flop to a primary output or the input of a flip-flop, but rather ends at a domino cell. For example, if a critical path consists of 20 identical cells in series, a balanced phase assignment would assign each phase 5 cell (in a four-phase clocking system). An unbalanced phase assignment could have many more than 10 domino cells assigned to a particular phase.

Risk evaluation. The risk associated with unbalanced phase assignments is in the maximum operating speed of the design. A faster process corner or a slower clock period obviously reduces this risk.

Unbalanced phase assignment, as described, is extremely unlikely to occur. This is because the phase assignment is based on a post-placed database for which accurate parasitic delay values are available. Each domino cell is assigned to a clock phase based on the last arriving data input rise time, with consideration made for expected clock skew. In addition, the skew-tolerant clocking methodology used can counteract phase assignment differences. For example, for a four-phase clocking system with a 50% duty cycle clock, an overlap window of 25% of the clock period allows phase distribution errors to be softened.

Problem detection. The problem is detected by the input data rising with respect to the clock falling setup check. This is shown in Figure 4.6.

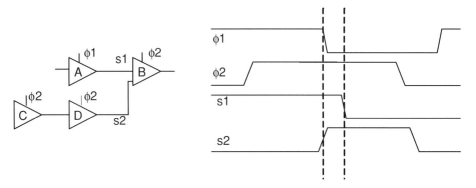

Figure 4.7. Fan-in phase differences and skew can lead to errors in domino cells.

Problem solution. While the phase-assignment algorithm should avoid this problem, it may emerge after routing. This is particularly true if many changes need to be done following the initial placement used in phase assignment.

4.4.3 Fan-in phase differences

After phase assignment a large number of cases occur in which two cells on different phases drive a logic cell. Thus, for example, cells clocked on phase ɸ1 and ɸ2 may drive a cell on phase ɸ2. This happens due to the reconvergence of a critical and a non-critical path, and is shown in Figure 4.7. The input clocked on phase ɸ2 may rise just before 50% of the clock period is complete. Since this is the same time at which ɸ1 falls, one has to ensure that the data clocked on phase ɸ2 rises before the data clocked on ɸ1 falls. Different fan-in phases can only cause errors if the two input-driven NMOS transistors are in the same serial stack, such as in an AND gate. The problem occurs if the interclock skew is greater than the precharge delay of the early arriving signal.

Risk evaluation. Faster processes reduce the likelihood of this problem. Let us explain why. For the case mentioned above, a faster process reduces the rising delay for the ɸ2 clocked input more than the falling delay from the ɸ1 clocked input. This follows since the rising input traverses through several cells (since its output rises approximately a quarter cycle after ɸ2 rises, it cannot be the first cell on ɸ2). The cumulative reduction in rising delay through several cells will be greater than the reduction in falling delay for a single cell (ignoring the pathological condition where the design is operated at a speed of four domino gate delays with each cell consuming a quarter of the clock period!). In addition, at a faster process corner the skew will probably be reduced. Finally, the input rising threshold is defined as the point at which the data reaches 40% of Vdd to be consistent with the static library. Domino gates tend to switch at lower input voltage values than static cells (as described in Chapter 1), so it is more likely to have switched by the time its input reaches 40%.

Running the design slower eliminates the problem since it gives the late rising signal more time to arrive.

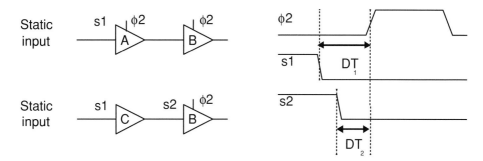

Figure 4.8. Reducing domino logic synchronization overhead by converting some domino logic cells to static logic cells.

For a domino design employing three clocking phases, each with a 50% duty cycle, no fan-in phase differences are allowable. For skew-tolerant domino methodologies employing five or more clock phases, the effects of fan-in phase differences are less constraining. This is because the earlier phase always falls when the latter phase is still active. Use of a longer duty cycle (at least 67% of the clock period) allows greater fan-in phase differences at domino cells.

Problem detection. This problem is checked using the minimum pulse width high overlap check that each domino cell is characterized for.

Problem solution. The minimum overlap must be satisfied for all domino cells. This can be done by reassigning the clock phase or inserting extra buffers on inputs driven by earlier clock phases. Unfortunately, because this check is not understood by the tool, an overlap error corrected during phase assignment may emerge again at the end of routing.

4.4.4 Static input ports

A static input signal to a domino module can change non-monotonically before settling down to its final value. After an input is stable it can be clocked on the rising edge of a clock phase. If the static input signal has different rise and fall delays then it must be clocked after the last possible transition to ensure that the domino cell is not erroneously discharged. The synchronization overhead with using static inputs can be up to the delay difference, plus the interclock skew, between two adjacent clock phases. For a four-phase clocking system the delay difference between two phases is 25% of the cycle time. For a three-phase clocking system it is 33% of the cycle time.

An obvious conclusion from this is that for critical paths all critical signals should be provided as domino inputs, i.e., inputs which always return to 0 before rising. This can be done with domino-compatible flip-flops or by some other scheme. If this is not done there will always be some delay overhead in providing static inputs to domino logic.

To reduce the overhead of dealing with static inputs it is possible to convert some domino cells to static ones. This will reduce the dead time overhead for synchronization. To understand this better we refer to Figure 4.8. For the implementation given at the top of the figure it can be seen that the static input signal on the wire s1 has a long dead time wait (DT_1) before being latched into domino cell A. At the bottom of Figure 4.8 it can be seen that by converting the first cell to a static cell, C, the time wasted for synchronization (DT_2) is reduced.

One must be careful to remember also that the clock signal in the module driving the static input port may have skew with the internal clock signals. This will cause the static signal to arrive early or late. Early arriving static signals do not stress the timing constraints for the design.

Risk evaluation. With faster transistors the speed of the static signal reaching the domino cell will be less. This is more likely to ensure that the static signal is stable before it is clocked. Slowing down the operating speed of the design has a similar effect.

Problem detection. This is checked in timing analysis, as the module will not be able to operate at the synthesized clock speed.

Problem solution. All static ports to a domino module need to be specified as "static ports" in the synthesis constraints file. The specified input delay must include the intermodule clock skew between the static driving module and the current one. If the static signal input delay is less than 75% of the clock cycle, the static input needs to be clocked on a domino phase once it is stable. Otherwise, the static input must be clocked at a flip-flop input (with the data only having traversed through static cells). The phase-assignment algorithm deals with all static input synchronization issues.

The phase assignment changes the domino cell clock phases to minimize the number of buffers added to avoid phase skipping. Such changes cannot be applied to domino cells receiving static inputs, whose phase, once assigned, must remain frozen. Any phase-skipping errors should be corrected by applying domino buffers to the inputs or outputs of the domino cell receiving the static inputs.

4.4.5 Domino input ports

The clock signal associated with a domino input port may be skewed with respect to the local clocks. Static input ports cause the input to arrive late in the worst case, after which the signal remains valid. For domino input ports the problem is more complex, because it could cause an input to rise late or to fall early. A maximum rise delay and a minimum fall delay are used to specify the domino input arrivals. These input delays must include the effects of intermodule skew.

Risk evaluation. Faster transistors reduce the possibility of overlap failures with domino cells since the cells switch quicker. Slower clock rates also reduce the possibility of failure.

Problem detection. Any domino cell receiving a domino primary input must ensure that a minimum overlap requirement for its input signals is met. There is no independent mechanism to check for this at the module level. Top-level timing analysis or dynamic simulations should validate that the specified rise and fall delays for all the input ports (static and domino) are correct.

Problem solution. The phase-assignment procedure should ensure that all domino inputs are correctly clocked.

4.4.6 Clock high width check

Assuming that the data signal is correct, the clock to the domino cells must be wide enough to enable the data to propagate through the domino cell.

Risk evaluation. If the input digital clock has an acceptable input duty cycle it is unlikely to be severely distorted in the clock tree. Faster silicon or slower clocks tend to reduce the possibility of pulse width degradation.

Problem detection. The minimum clock pulse high width constraints associated with each domino cell are used to test for this condition.

Problem correction. No problem correction for this failure mechanism is part of the standard flow. This is because the possibility of this type of error occurring is negligible. Depending on the cause of the failure, the clock tree synthesis can be rerun or a different clock source used. Since the clock source is often provided from an external module to the domino circuit, top-level analysis should be used to validate the specified clock duty cycle and transition times.

4.4.7 Precharge failure

The errors we have discussed till now have been evaluate phase errors that can arise from a straightforward application of the phase-assignment procedure. In addition to these problems, precharge errors can also be manifest after phase assignment. The most direct manifestation of a precharge failure is if the precharge time for a cell is insufficient. This should only happen if the design is run above the maximum operating frequency or if the intraclock skew is terrible.

Risk evaluation. The precharge time and skew is less with a faster process corner. Also, with more cycle time the precharge has greater time to complete. Thus, for faster processes the danger decreases for this failure mechanism.

Since precharge delays are always generally less than the precharge period, this error is unlikely to occur.

Problem detection. The cells non-sequential setup rising check detailed in Chapter 3 detects this problem.

Problem correction. The clock tree synthesis procedure will produce a clock tree with an acceptable skew. If the skew is worse than budgeted, the clock tree synthesis will need to be rerun. A systemic precharge failure may indicate that the domino library is designed with insufficient precharge delays.

4.4.8 Static output ports

The domino logic precharge data must not be transmitted into the static logic as it can lead to functional errors.

Risk evaluation. The possibility of a precharge value propagating out of the module increases with faster transistors. The possibility of this failure occurring is independent of operating frequency.

Problem detection. The "static output" ports are specified in the synthesis constraints file. During signoff timing analysis every static output port should be checked to ensure that the clock skew (between the output driving cell and the latch) is less than the minimum precharge path delay (the total precharge delay plus the latch setup). This should be done at both the fast and the slow corner.

Problem solution. A domino cell on phase clk, ϕ1, ϕ2, and ϕ3, whose output drives a static module, must go through a latch which is clocked with the same phase. The latch must be transparent when the clock phase is in evaluate and opaque when it is in precharge. If the static design is clocked with the same clock as the domino design (clk, which has the same phase as ϕ1), then the phase ϕ4 does not need a latch placed after it. In domino synthesis we use separate clk and ϕ1 clock since the loading is significantly higher on the ϕ1 clock, leading to greater interclock skew. Since the clk signal is used for hard edges, a tighter transition time and skew tolerance is possible when separated for the ϕ1 phase.

Precharge hold failures can be corrected by adding extra static buffers between the domino cell and its latch (which is the standard hold time failure fix). Alternately, if the intraclock skew is large it may be tightened to correct this problem.

4.4.9 Clock width low check

This problem will lead to a precharge failure. The clock pulse low width check is used to check this topic.

4.4.10 Multi cycle paths

Domino logic designs may include multi cycle modules. Common examples of such modules are multipliers and dividers. The phase and timing rules specified for domino logic cells continue to apply in multi cycle designs, with only two changes required in the design flow.

The first change relates to phase relationships across register boundaries. For example, it may be possible that in a two-cycle module a non-critical path traverses through five cells clocked on clock phases ϕ1, ϕ2, ϕ3, and two on ϕ4. The first cell on ϕ4 may be in the first cycle and the second cell in the second cycle. A straightforward adherence to clock phase relations across the two cycles would require the insertion of clock buffers on ϕ1, ϕ2, and ϕ3 between the two cells on phase ϕ4. This may not, however, be necessary as the second cell on phase ϕ4 may receive its input signal early from the first cell on ϕ4. What must be ensured is that if a non-critical input to a cell skips cycles (not phases), then this does not lead to a new critical delay in the design. For example, if a domino design has a 4 ns clock cycle, then adding an extra cell in the last phase is acceptable provided that the total cycle time of the design does not exceed 4 ns. This condition ensures that timing constraints are satisfied in the multi cycle path. Since the output will arrive early, however, this may require changes to the logic that interact with the multi cycle path output. If this is to be avoided, extra domino buffers clocked by the appropriate phases will need to be added.

The second change required for multi cycle paths relates to how they must be specified. Experiments with a timing analysis tool suggest that the non-sequential setup time violations are flagged only when multi cycle paths are defined between different clocks. This means that a different number of multi cycle stages need to be defined between clock phases in different stages. For example, a multi cycle path of 2 must be set between cells on phases $\phi1$ and $\phi4$ when they are both in the first cycle. When a cell on phase $\phi1$ is in the first cycle, and another cell on phase $\phi4$ is in the second cycle, then a multi cycle path of 3 must be set between them. If the phases are not differentiated between consecutive pipeline stages, then multi cycle path constraints will have to be defined explicitly between gates belonging to different stages but sharing the same phase names. This can lead to a huge number of gate timing exceptions (multi cycle paths) that will have to be added for the timer to understand correctly the behavior of the pipelined logic. Even if this process could be automated during the phase-assignment process, the resulting additional constraints would add timing complexity and slow down the EDA tools during the rest of the optimization process. The preferred solution we used was to duplicate the phase names across pipeline stages. For instance, phases $\phi1$, $\phi2$, $\phi3$, and $\phi4$ of the second stage of the pipeline can be renamed $\phi5$, $\phi6$, $\phi7$, and $\phi8$. The relationship between the elements of different stages then becomes easier to describe. One can in fact describe the delay between $\phi1$ and $\phi5$ as a clock (source) latency, suppressing the need for any additional timing exceptions, or with the clock waveform definition available in synthesis tools.

4.4.11 Phase assignment with static and domino cells

Before proceeding let us review the topics discussed till now. We saw that starting from an initial phase assignment the only potential changes that need to be made are for phase skipping, fan-in phase differences, or static ports. Of these changes, phase skipping and fan-in phase differences should not impose any direct delay penalty. This is because they represent problems caused by early arriving signals, which can be delayed with no timing penalty. Static input ports can impose a more serious penalty, due to the signal having to wait for a synchronizing clock edge. This overhead can be minimized by adjusting the synchronizing clock arrival (useful skew) in order to avoid having the data wait for the capturing edge of the clock. The flow to accomplish this is, however, complex. It is easiest to solve the problem by changing the static inputs into domino-compatible ones (with special flip-flops). These changes will require effort to verify the new RTL code. We believe that this effort is worthwhile in high-speed design where domino logic will be applied. Static output ports also have an overhead due to the extra latch. This can be minimized by incorporating the latch into the domino cell if it is on a timing-critical path.

The phase-assignment problem for designs employing only domino cells is thus well understood. The process becomes more difficult when static cells are intermingled with domino ones. The advantage of using static cells, however, justifies the greater complexity. These advantages include substantially lower power dissipation and a more routable design (a major advantage if the routing tool can intelligently reroute wires to reduce crosstalk). The area of the design is also reduced. The static cells are likely to be introduced into the design through a series of incremental compiles after the design has been unated or placed. The incremental compile should not increase the delay in the design. Since the static cells will be used to replace domino cells, only non-inverting static cells will be provided (while a non-inverting domino cell may be replaced by two inverting static cells, adding inverting logic may also lead to other, invalid logic transformations).

With the static and domino cells present, the initial phase assignment is the same. All domino cells are assigned an initial phase based on the last arriving input to the cell. The problems that can occur now are extensions of the problems that we saw when only domino cells were present. For example, instead of phase skipping occurring when a domino cell on ϕ1 drives a domino cell on ϕ3, this could now happen through a static buffer. Fan-ins of different domino clock phases to a static cell need to be analyzed for the more stringent minimum pulse width high check. Static input ports may traverse through several static cells before being latched into a domino cell. When static cells are added to domino logic modules, they tend to occur mostly on non-critical paths.

During phase assignment some static cells will need to be converted back to domino cells to ensure that no evaluate or precharge failures occur. Converting static cells to domino cells can be done by a direct substitution of the static cells with an equivalent domino one (in terms of drive strength and input capacitance). This should not alter the maximum operating speed of the circuit, since static cells are slower than domino ones. Once a static cell has been converted to a domino cell, care must be taken to ensure that it is not reverted to a static cell in any subsequent optimization. Setting "don't_touch" attributes on the domino cells will ensure this. Having static gates between domino cells does reduce the likelihood of errors due to phase skipping, as phase skipping is similar to a hold failure, with the extra static gates acting as delay elements. Under such circumstances, phase skipping may be acceptable in domino cells.

4.5 An example domino synthesis flow

We have described the unate and phase-assignment processes needed in domino synthesis. In this section we describe a synthesizable domino logic RTL to GDS flow designed for a 130 nm CMOS process that was subsequently refined for a 90 nm process. The flow is based on an existing static flow, with modifications being made, where needed, to support domino logic synthesis. While this flow description is specific to a set of tools, depending on the availability of EDA tools the domino synthesis flow can be modified. This is a similar situation to static logic, where flows are modified as tools and features improve.

4.5.1 Overview

Figure 4.2 shows the steps in the domino logic flow. The flow starts with a standard static RTL synthesis. This step uses Synopsys's Design Compiler™. The design is subsequently unated by applying bubble pushing. This is also done within Design Compiler™. Since logic is changed during the unate step, an incremental compile optimization is subsequently performed.

The domino logic block is then placed. This step is done with Synopsys's Physical Compiler™. Based on the actual arrival times on the input data pins, clock phases are next assigned to the domino cells. This analysis is done using PrimeTime™, with the required netlist modifications (cell replacement, buffer insertion) executed by means of an incremental optimization in Synopsys's Physical Compiler™. After placement, the clock tree synthesis and routing is performed. For this too a Synopsys tool is used: Astro™. Clock tree synthesis is also done with Astro™. The subsequent routing and optimization is timing-driven and crosstalk-aware.

The initial synthesis and bubble pushing is done using the pseudo-static version of the domino library. From phase assignment and onwards, the flow switches to the complete domino library which includes the clock pins.

4.5.2 Domino flow-specific variables

As in most RTL to GDS flows, the domino synthesis flow requires a number of variables to be defined. These variables define the modules being synthesized, the expected input and output files produced during the flow, synthesis constraints, and tool settings. In addition, a number of other variables are used to define domino-specific variables. To get a better flavor of the flow, we provide a short description of their purpose. Further elucidation for the purpose of these variables is provided in the sections that follow. We start by describing the general or overview variables:

dom_worst_case/dom_best_case: Defines the best- and worst-case corner to be used for synthesis. Since the worst-case crosstalk corner may not be the same as the worst-case delay corner, this variable may be varied later to check crosstalk susceptibility of the design. This is similar to a static flow where two or more corners are defined.

dom_pt_analysis_type: Specifies whether the timing analysis should be for a single case, or use worst-case and best-case corners. Again, this is similar to a static synthesis flow.

There are some template synthesis script variables that are used in domino synthesis. They include:

dom_input_delay: Default input delay on the data input ports (in nanoseconds). This delay is specified with respect to a clock.

dom_output_delay: Default output delay on the output ports (in nanoseconds). As in the input delay, the output delay is with respect to a clock.

dom_output_load: Default load on the output ports (in picofarads).

dom_input_transition: Default input transition on the input ports (in nanoseconds). Transition time impacts the delay directly through a gate, so as good practice it should be specified.

Domino logic uses more clocks than static synthesis. A number of variables are needed to define the clock variables:

dom_clock_prefix: The prefix used for describing the different clock phase names. The default name is phi, represented by the Greek symbol ϕ. Using this prefix simplifies searching the database for the clock signals for domino cells.

dom_main_clock: The name of the clock that triggers the domino flip-flops. If there are no flip-flops in the block this variable should be left empty. The default name used is clk. Having a separate clock for the flip-flops ensures that they can be assigned a tighter clock skew than the domino clocks. This is useful since the flip-flops have hard clock edges, whereas domino cells use softer clock edges. If all clocks are specified with a tight skew bound, the tool would attempt to use very large buffers in the clock tree. This is unnecessary, since the domino cells are skew-tolerant, leading to excessive power dissipation.

dom_period: This defines the target clock period.

dom_sys_clock: A virtual clock name used as a reference for the static input and output delays.

dom_phase_number: The default is four clock phases. More or less clock phases can be used as desired. How many clock phases will be used depends on what clock sources are available.

dom_cycle_number: The number of clock cycles to compute the output result before it is registered. This variable is used in modules with multi cycle paths. The default value of this variable is 1.

dom_clock_min_transition: Minimum clock transition for the clocks.

dom_clock_max_transition: Maximum clock transition for the clocks.

dom_clock_min_uncertainty: Minimum clock uncertainty for the clocks.

dom_clock_max_uncertainty: Maximum clock uncertainty for the clocks.

dom_clock_min_insertion: The target for minimum clock insertion delay to be used during clock tree synthesis.

dom_clock_max_insertion: The targeted maximum clock insertion delay. The clock insertion delay is important when a domino module interacts with logic from other blocks.

The domino synthesis needs to control the synthesis and placement options used:

dom_incr_compile_runs: The total number of incremental compiles after bubble pushing. The default value is 2.

dom_incr_add_static: Defines at which incremental compile iteration the static cells should start being allowed. The default value of 0 does not allow static cells to be inserted before placement.

dom_place_add_static: A value of 1 allows static cells to be used on non-critical paths during placement. The default value of 0 does not allow static cells to be inserted during placement.

dom_physoptInitialMode: Variable to control Physical Compiler™ settings. Using timing-driven congestion, which is the default setting, will balance the quality of the placement between timing and congestion metrics. Using only congestion-driven placement tends to penalize timing. It is suggested to use the additional option of "-congestion_effort high" if timing-driven congestion is chosen.

dom_physoptIncrementalMode: Using the default setting of "-congestion" invokes additional checks during incremental placement to prevent routing congestion.

dom_physoptInitialEffort: The default is medium. Changing it to high may increase runtime and disables the "-congestion_effort switch". The quality of results is generally better with high effort.

dom_physoptIncrementalEffort: This is the effort level provided during Physical Compiler™ incremental runs.

As in static synthesis, target libraries need to be specified for domino synthesis:

dom_static_core_libname: List of the static target libraries to be used during synthesis and optimization.

dom_domino_libname: List of the domino target libraries to be used during synthesis and optimization.

The unate or bubble-pushing step requires a number of variables to be defined:

dom_static_ports: By default the flow assumes that all the input and output (I/O) ports, with the exception of the clock ports, are domino signals. This variable is used to define the I/O ports that come from a static gate or that will drive a static output (do not put the clock pins or the special static ports in this list).

dom_special_static_ports: List of the ports that drive the scan and reset pins of the domino registers. The fan-out cone of these ports cannot include any domino cells.

dom_max_fanout: This is the threshold above which an explicit inverter is replaced by a buffer cell during the unate operation. The default value of the variable is 2.

dom_std_inverter: Specifies the inverter to be used when creating the dual version of a static input port. This variable points to an inverter in the available static library.

dom_std_buffer: Specifies the domino buffer used in replacement of an inverter. This works in conjunction with dom_max_fanout.

dom_output_latch: Specifies the static latch that is to be inserted before a static primary output.

dom_output_latch_F: Same as above except when the output signal has to be inverted.

The next set of variables control cell placement:

dom_HFNFanoutThreshold: Fan-out above which a net is marked as a high fan-out node (HFN). The default value is 80. HFNs are set ideal during Physical Compiler™ optimization. They are synthesized in Astro™, the final router.

dom_dffs_opt: The standard flow uses dual-output flip-flops for synthesized registers. Enabling this optimization, which is the default case, replaces dual-output D-type flip-flops (DFF) with single-output DFFs when only one output is used.

dom_output_phase_opt: Enables the output phase inversion optimization technique to reduce the area of the unated design. By default it is off, as this optimization is very time-consuming.

dom_keep_xor_static: It is possible to allow static XOR or XNOR cells if they are located just before registers or the static primary outputs. This avoids requiring the fan-in cone of the XOR to be duplicated. The default value for this variable is false, since this rarely occurred in the relatively complex designs we worked on. Using a static XOR cell may also have a speed penalty compared with a domino implementation.

dom_dual_prefix: Prefix used to create the dual version of a net or of an input port. The default value is "F_".

The phase-assignment algorithm works on placed designs. A number of variables need to be specified for phase assignment:

dom_default_corner: The operating corner used during phase assignment. The default value is worst.

dom_pt_analysis_type: Analysis type during phase assignment. The default value is worst case and best case.

dom_skip_limit: Limits by how much a gate phase can be shifted to fix a phase skipping. The default value is 1, which means no phase skipping can occur.

dom_skip_if_slack: This variable controls the prevention of phase skipping based on the slack available at an endpoint. The purpose of this variable is to ensure that an extra buffer is not inserted if slack does not exist at the endpoint. The default value is -9999, which means that phase skipping is avoided before a flip-flop.

dom_slack_tolerance: Aborts phase assignment if the worst negative slack (WNS) is less than the specified value. The default value of the variable is negative 0.02 ns. As in static synthesis, the design should be constrained so that the clock period can be met.

dom_preconnect: During phase-assignment iterations, previously inserted buffers are automatically connected according to whether their previous phase assignment in this variable is true. This is the default setting. This variable is needed to ensure that the phase-assignment process uses the results from an earlier iteration.

dom_skew_clocks: Used to delay the clock rather than completely skipping a phase to handle static to domino interfaces. Default value is true. For instance, if a static input feeding a domino gate becomes stable shortly after the rising edge of $\phi 2$, rather than waiting for the rising edge of $\phi 3$, the capturing domino gate will still be clocked by a delayed version of $\phi 2$. An additional constraint will be added to the clock connection to inform the clock tree synthesis tool that this clock signal has to be delayed (skewed) with respect to the other clock pins assigned to the same phase.

dom_ignore_inactive_pins: The default value of this variable is false. This variable ensures that during phase assignment, if a cell is encountered where some timing arcs are disabled, then the last arriving input timing will not consider the disabled pins. For example, if pin A of a cell implementing the function (A*B + C) is set to zero, then the timing arc from B to Z is ignored. One is unlikely to encounter such a situation in a well-optimized design.

dom_simplify_latched_outputs: If true, the output latches that are clocked by the last phase are replaced by a buffer or an inverter. This is possible when the static logic will be registered at the input of the next block. The default value is false.

dom_skip_if_static_driver: Allows phase skipping if at least one static cell is present between the two domino cells. The static cell makes the likelihood of a hold failure less.

dom_use_mixed_reg: If a domino register drives a cone of logic that only has static cells and a cone of logic that has domino cells, it is possible to support both logic families with a mixed register that has both domino and static outputs. The default value of this variable is true.

dom_phase_use_sbf: Allows static delay cells in place of domino buffers to fix phase-skipping errors. The default value is true.

dom_sorted_buf_list: List of domino buffers used to fix phase skipping. The list is sorted by the capacitance the cell can drive. This eases the initial cell substitution process.

skw_sorted_buf_list: List of static delay cells used to fix phase skipping. Skewed buffers are special static buffers used to correct phase skipping. They have fast rise delays and slow fall delays to ensure that delaying the precharge does not slow down the evaluate path excessively.

A last set of variables should normally not be modified by the user. They are used when parsing the netlist for easy identification of cell types:

dom_static_inv_prefix: Prefix used to identify inverters in the netlist. The default value is "IV".

dom_static_lat_prefix: The prefix used to identify the static latches. This is very library-dependent. In our case, the default is "LD".

dom_prefix: The prefix used for all domino cells. The default value is "dom_".

dom_buf_prefix: The prefix used to identify domino buffers. The default value is "dom_BF".

4.5.3 Design guidelines

Although the domino library comes with high-performance flip-flops, they do not need to be used. For example, one can advantageously replace a multi cycle pipelined design having explicit flip-flops, with domino cells and appropriate clocking. For this a multi cycle constraint needs to be specified. For the domino logic flow one may, however, want to use flip-flops if:

- Any existing RTL model is to be used without modification.
- Scan needs to be supported.
- The data flow has feedback loops.

It is always general good practice in RTL descriptions to systematically latch the inputs and outputs of logic blocks. If the primary output of a domino block drives only static gates, then the phase-assignment algorithm will try to reduce area overhead due to logic duplication by not providing the signal and its dual. If an arbitrary block in the chip sends data to a domino block whose inputs are not registered, it is best to use domino flip-flops

Table 4.1. Constraints used in domino synthesis.

Constraint type	Pseudo-static	Full domino
Clock definition	System and main clocks only specified with: **create_clock -period**	Domino phases along with the system and main clock. Clock latency also specified: **create_clock -period** **set_clock_latency -source**
Input/output delays	**set_input_delay** **set_output_delay**	**set_input_delay -max -rise** **set_input_delay -min -rise** **set_input_delay -max -fall -clock_fall** **set_input_delay -min -fall -clock_fall** **set_output_delay -max** **set_output_delay -min**
Boundary constraints	**set_driving_cell** **set_load** **set_max_capacitance**	**set_driving_cell -max/-min** **set_load -max/-min** **set_max_capacitance**
Clock constraints	**set_clock_transition** **set_clock_uncertainty**	**set_clock_transition -max/-min** **set_clock_uncertainty -max/-min**
Timing exceptions	**set_multicycle_path** **set_false_path**	Additionnal constraints added by the phase assignment script

for this interface. These flip-flops can produce both the output and its complement as a domino-compatible signal.

When a domino cell is driven by a static signal, the phase assignment will have to ensure the clock that drives this cell rises at least a setup time after the worst fall arrival of the signal. This will lead to a delay penalty. Therefore, critical static delays should not directly feed a domino combinational block.

4.5.4 Constraint settings

The domino flow requires a minimum set of constraints to be defined to operate properly. These constraints fall into five categories:

- Clock definitions.
- Input/output delays.
- Boundary conditions.
- Clock constraints.
- Timing exceptions.

Since the flow operates in two different modes (pseudo-static from RTL to placement and full domino from phase assignment to GDS), there will be two sets of constraints. The full domino constraints will only be used during place and route, and signoff. Table 4.1 shows how the different phases should be set for pseudo-static and full domino modes.

Initially only the register clock, unless the block is without flip-flops, and the virtual system clock need to be defined. Examples of this are shown below in the Synopsys design constraints (SDC) format:

create_clock -period 2.0 -name clk [get_ports clk]
create_clock -period 2.0 -name sysclk

During phase assignment, the domino phases need to be added. This is done automatically. Here is an example of clock definitions added for the following domino variables:

dom_phase_number = 4
dom_cycle_number = 1
dom_clock_prefix = phi
dom_period = 2

The corresponding commands are:

create_clock -period 2.0 [get_port phi1]
create_clock -period 2.0 [get_port phi2]
create_clock -period 2.0 [get_port phi3]
create_clock -period 2.0 [get_port phi4]
set_clock_latency -source 0.5 phi2
set_clock_latency -source 1.0 phi3
set_clock_latency -source 1.5 phi4

One should note that if domino synthesis uses N clock phases, then clocks from "phi1" to "phi$N$" will be created. The source latency for phi$N$ will be $(N-1)/N$ times the clock period.

When pushing a design for speed, it may help to slightly overconstrain the design. One must be careful, however, as excessively overconstraining the design will lead to sub-optimal results and very long runtimes. If synthesis leads to large negative slack (-20 ps is the default threshold set in the flow), one will need to relax the clock period to bring back the worst negative slack (WNS) above the threshold. Failure to do this will cause phase assignment to abort. A low WNS also improves the cell area and cell utilization in the design.

Accurate input and output delays are needed to allow the current block to interface correctly with the rest of the design. Input delays on static inputs are particularly important since they directly influence the phase assignment. The phase that will capture a static input has to arrive after the input is stable. If an input is critical, it should be a domino input, i.e., latched with a domino flip-flop. Remember that domino I/O ports can never be bidirectional. Some guidelines for specifying input and output delays:

- Static input and output delay constraints should be set with respect to the virtual system clock (dom_sys_clock variable). The full set of port constraints required for the phase assignment should show four constraints, in order to model separately the rising and falling transition in the best and worst corner. For example, let us consider that CTRL is a static signal coming into a domino block. The rise and fall delay for the signal are at 300 ps and 280 ps in the worst case, and 150 ps and 140 ps in the best case. The following constraint should be defined for phase assignment:

set_input_delay -max -rise -clock dom_sys_clock 0.300 [get_ports CTRL]
set_input_delay -max -fall -clock dom_sys_clock 0.280 [get_ports CTRL]
set_input_delay -min -rise -clock dom_sys_clock 0.150 [get_ports CTRL]
set_input_delay -min -fall -clock dom_sys_clock 0.140 [get_ports CTRL]

Since the pseudo-static library cells have equal rising and falling timing arcs, a single input and output delay is enough prior to phase assignment. When using the psuedo-static library, one will need to anticipate the time penalty needed to synchronize static inputs with the domino circuitry. Depending on the synchronization strategy, the input delay should be increased to account for the timing penalty. For example:

set_input_delay -clock dom_sys_clock 0.580 [get_ports CTRL]

- Domino input and output delay constraints must be set with respect to an actual domino clock phase. As with static inputs, four constraints are required for the phase assignment to model separately the rising and falling transition over the best and worst corner.
- For domino inputs the rising arrival should be defined with respect to the rising edge of a clock phase and the falling arrival should be defined with respect to the falling edge of the same clock phase. For example, let us consider a domino input coming directly from a domino flip-flop DIN. The rise and fall delay for the signal is 100 ps and 280 ps under worst conditions, and 50 ps and 140 ps under best conditions. The following constraint should be defined for the phase assignment:

set_input_delay -max -rise -clock clk_phase1 0.100 [get_ports DIN]
set_input_delay -max -fall -clock_fall -clock clk_phase1 0.280 [get_ports DIN]
set_input_delay -min -rise -clock clk_phase1 0.50 [get_ports DIN]
set_input_delay -min -fall -clock_fall -clock clk_phase1 0.140 [get_ports DIN]

During initial synthesis with the worst evaluate, timing only needs to be specified. That can be done as follows:

set_input_delay -clock clk_phase1 0.100 [get_ports DIN]

It is essential to specify accurate boundary constraints for the I/Os. The boundary constraints needed in SDC format are given below:

- A driving cell and a load for each input port (including the clock ports)

**set_driving_cell -library library_name.db -lib_cell BFSVTX8 **
[get_ports INP]
set_load 0.1 [get_ports INP]

- A maximum capacitance on each input port

set_max_capacitance 0.08 [get_ports INP]

- A load for each output port

set_load -pin_load 0.1 [get_ports OUTP]

The only clocks that have to be constrained up-front by the designer are the main clock used to trigger the registers and the virtual system clock used as a reference for the input and output delays. The domino phases are created and constrained automatically during phase assignment. The clock constraints needed are:

- A transition for each clock

    set_clock_transition 0.15 [get_clocks clk]

- An uncertainty to account for process variations and skew

    set_clock_uncertainty 0.05 -setup clk
    set_clock_uncertainty 0.1 -hold clk

- Network latencies to model clock insertion delay

    set_clock_latency 1.2 sysclks

Any timing exception required in the design will need to be specified with these constraints. The reader is reminded that exceptions should be used with extreme care as they can invalidate timing analysis results and hide critical paths. Examples of timing exceptions are multi cycle paths and false paths. During phase assignment, additional exceptions will be added automatically by the script to account for the domino clocking scheme.

4.5.5 RTL description

For domino synthesis it is possible to reuse the exact RTL used for a static implementation. Be aware, however, that domino registers only have a synchronous reset. Ideally, you would like to describe your registers synchronously. An example of a synchronously reset register is given below using the Verilog hardware description language:

reg PM;
// synopsys sync_set_reset "resetn"
always @(posedge clk)
  if (!resetn)
    PM <= 0;
  else
    PM <= data;

If the RTL uses asynchronous flip-flops which cannot easily be changed, synthesis can still continue by using a set of dummy flip-flops defined in the domino library. These flip-flops do not have physical views. Their timing is copied from their synchronous flip-flop equivalents. To ensure that these dummy flip-flops are not inadvertently used, they have the "don't_use" attribute assigned to them. During bubble pushing these flip-flops will be substituted with actual synchronous flip-flops. If dummy asynchronous flip-flops are replaced by synchronous flip-flops, formal verification of the final netlist cannot be compared with the RTL used for synthesis. A comparison must instead be made with the synthesized netlist. The input RTL has to be compared with the final netlist using dynamic simulations.

4.5.6 Synthesis options

It is possible to reuse many of the synthesis options developed for a static design during domino synthesis. To reuse the synthesis options they must comply with the requirements listed. During bubble pushing the only domino cells that can be used are in the domino pseudo-static library. After bubble pushing the only cells that are allowed are the true domino cells (no inverting logic, dummy XOR/XNOR cells, or dummy asynchronously resettable flip-flops are acceptable). Skewed buffers should also not be present at the end of bubble pushing, as these cells are best introduced during phase assignment.

After bubble pushing the physical implementation process starts with placement. During this stage we do not wish logic transformations to occur which will add inverters to the design. To ensure this, the static library inverters are not allowed following bubble pushing. Since the physical synthesis tool may require the presence of an inverter in the library, a dummy inverter is provided. This inverter has a large delay and area number, ensuring that the tool will never use it in an optimization step. In addition, all inverters already present in the design should be marked as "don't_touch". It is recommended that the scan chains be stitched after placement and the high fan-out asynchronous network (the set, reset, scan_enable signals) be set to ideal. As long as one is compliant with these requirements, it is possible to use an existing synthesis script. A set of recommended template scripts is also provided.

4.5.7 Bubble pushing

As described previously, the purpose of the unate or bubble-pushing algorithm is to make the initial synthesized logic non-inverting. This is done by pushing the inverters toward the inputs using De Morgan's theorem. The domino synthesis algorithm allows for certain optimization steps to be controlled. Setting the variable dom_output_phase_opt to true, allows the bubble-pushing algorithm to try and remove trapped inverters by propagating the inversion toward the output. This technique cannot, however, fix the logic duplication due to the presence of XOR and XNOR cells. The variable dom_output_phase_opt should only be allowed when the duplication due to XOR and XNOR cells is below 60% of the logic and the number of endpoints for a logic module is less than 100 (an endpoint is either a register or a primary output). Having excess duplication makes output phase optimization unlikely to lead to improvements, while having more than 100 endpoints can lead to excessively long runtimes. Output phase inversion can only lead to a possible error if the output phase drives another domino module, without being latched initially into a register. In our flow all domino output signals are registered.

Every port that drives, or is driven by, a static gate should be specified with the dom_static_ports variable. When an output port is declared as static, a latch is automatically inserted in order to make sure that the output retains its value until the end of the cycle. This step occurs during bubble pushing, with correct clock phase being assigned to the clock pin during phase assignment. Until then the clock pin of the latch is set to one.

If the dual copy of an input signal is needed due to logic duplication, there are two possible outcomes. If the input signal is declared as static, then the dual signal is created from the input by inserting an inverter. If the input signal is declared as dynamic, then the dual signal has to be provided externally and therefore a new dynamic port is created (if it does not already exist), whose name derives from the original input attached to the specified dom_dual_prefix variable.

The bubble-pushing algorithm starts with some sanity checks. These checks ensure that all outputs and flip-flops have either domino or static endpoint attributes and that all domino flip-flops have dual outputs available. If a cell does not have a load, it is removed from the netlist. Subsequently, all inverters not flagged with the "don't_touch" attribute are removed, with the input net entering the inverter being tagged as "inverted". If the inverter being removed has a fan-out greater than dom_max_fanout, a domino buffer (specified with the variable dom_std_buffer) will drive the node. The input pin of the buffer will then be marked as inverted. This ensures that the inverter is replaced by a cell of sufficient drive strength. The fan-in cones of all XOR/XNOR cells are then marked for duplication. Next, starting at all endpoints, the logic is traversed to all the logic starting points (register outputs and primary inputs) with De Morgan's law applied to push all the inverters back. If dom_output_phase_opt is set to true, the phase optimization technique is activated. After checking that no inversions are left in the netlist, the final data and reports are written. The data is stored in the Synopsys.db format and as a Verilog netlist.

4.5.8 Post-bubble pushing incremental optimization

Due to logic duplication, bubble pushing may noticeably disturb the design. The variable dom_incr_compile_runs specifies the number of incremental optimization runs that are used to again achieve timing closure. A slight decrease in area, accompanied by a greater reduction in power consumption, may be possible by replacing domino cells on non-critical paths with regular static gates. This is done by allowing static cells to be inserted during the incremental compile iterations. The dom_incr_add_static variable indicates during which iteration the static cells start to be used in replacement of domino gates. If this variable is set to 0 or to a value greater than dom_incr_compile_runs, then no replacement is performed. That value must be in the range from 1 to the number defined by dom_incr_compile_runs in order for the static cell insertion procedure to be invoked. The primary power benefit of static cells is in their lower clock tree loading. If you choose to insert static gates and wish to keep them in the placement stage, the variable dom_place_add_static is set to 1.

The phase-assignment flow supports directly unating a gate-level netlist via the dom_db_ext command. This is useful if an optimized gate-level netlist is already available. Sometimes implementation groups are only provided with gate-level netlists to limit the distribution of proprietary RTL knowledge.

4.5.9 Initial placement

Physical Compiler™ is used to perform the detailed placement of the unated domino block. This operation will start from the latest Synopsys database produced by bubble pushing. A minimal floorplan can be specified in one of two ways: using a PDEF 3.0 file or using the minimum physical constraints (MPC) format supported in Physical Compiler™.

If the MPCs are used, the script will automatically generate a floorplan based on the defined Y/X ratio utilization. Placement blockages can also be placed on either side of the block. These blockages are strongly recommended, since they reserve space for antenna diodes. These protection diodes are inserted during routing. Along with the basic floorplan settings, MPCs can also be used to define the minimum spacing between ports. If the design has port placement requirements, the set_port_location or set_mpc_port_option variables are used to define the absolute or relative port placement or side constraints.

Depending on the accuracy of the wire load models (WLM) used, the timing and area results after placement may differ slightly from the results after bubble pushing. In some designs we have seen a speed degradation of up to 15%. It is for this reason that we follow bubble pushing with an incremental optimization step. The timing slack available on a placed design can be used to replace domino cells with static cells. Since the timing information is far more accurate after placement, this optimization can lead to different results from those achieved for the optimization done after the unate operation. Setting the variable dom_place_add_static to 1 (the default is 0) will allow static cells to be used during placement.

At the end of the placement the worst negative slack for the design should be above the threshold defined by dom_slack_tolerance. In order to ensure that the timing will not be significantly worse after routing, wiring capacitance is derated by 15% during placement, with actual power grids pre-placed. This allows the placer to correctly estimate the routing tracks required by the power and ground grids. In order for the placement to be routable, the final utilization should remain below 80%. One should also check the congestion report and look at the violations of the congestion thresholds (the number, the average values, and the deviations therein).

4.5.10 Phase-assignment variable options

Up to this point the domino block has been used with a pseudo-static library model. This model has accurate performance, power, and area data for the domino standard cells but no clock pin. This approach allows the use of traditional synthesis and placement

Domino logic synthesis 99

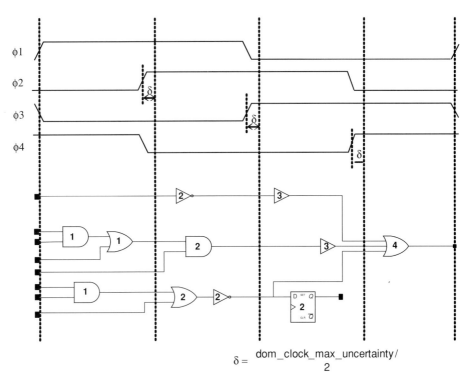

$$\delta = \frac{\text{dom\_clock\_max\_uncertainty}}{2}$$

Figure 4.9. Phase assignment with four clock phases.

optimization techniques available in Design Compiler™ and Physical Compiler™ or other synthesis and physical design tools.

In the phase-assignment algorithm, the pseudo-static cells are mapped to regular domino cells and the clock of each cell is connected to a particular clock phase. The current flow is optimized to support four overlapping 50% duty cycle clock phases. The clock phases are evenly split, i.e., they have a 90 degree phase shift with respect to each other.

The first domino clock phase, $\phi 1$, and the register clock signal are treated as independent ports in the module, although they will likely be tied to the same source outside the block. The separation in the blocks allows for a tighter clock skew bound to be applied to the register clock. The bound on interclock skew between the domino clock phases is critically important for phase assignment to be valid. The faster the block, the tighter the clock skew requirements. For 130 nm designs, for example, a 1 GHz clock should have an interclock skew below 100 ps.

A graphical illustration of the phase-assignment strategy is shown in Figure 4.9. Four clock phases are used, with each domino gate assigned to a window according to the latest arrival time on its data inputs. Domino cells on the boundaries of two clock phases can slow down the design if the cell is placed on the latter phase and this phase is delayed due to clock skew. The possibility of this is minimized by adding a margin, δ, which is half the expected clock skew, by which time all arrival times are shifted to an earlier clock phase. A more detailed description of some of the choices that need to be made by the phase-assignment procedure is given next.

If the input pin of a domino gate is driven by a static signal, the clock input should rise after the static data has fallen. This also means that the most critical path through this gate will now start from the clock pin. To handle this situation the dom_skew_clocks variable is considered. If dom_skew_clocks is set to false, the script will assign the gate to the first phase that can safely capture the data. This can cause up to a quarter of the clock period to be wasted. If dom_skew_clocks is set to true, the script will calculate the earliest needed arrival time, T, for the clock. If the clock insertion time is large (the minimum clock insertion time is set to 100 ps), the next clock phase will be used to clock the gate, and a phase shift constraint will be set on the clock pin to advance the clock arrival to time T (negative skewing). If it is not the case, then the gate will be clocked with the earlier phase and a phase shift constraint will be set on the clock pin to delay the clock arrival to time T (positive skewing).

When a primary output has been declared static (meaning that it will drive static gates), a latch is inserted during bubble pushing. To determine what domino phase should trigger the latch, the script will look for the phase that generates a falling transition (precharge) on the input of the latch and assign the latch to this same phase. If it cannot find any, the clock pin of the latch will be tied to Vdd. Setting dom_simplify_latched_outputs to true removes latches that are clocked by the last phase and replaces them with a buffer or an inverter.

If a domino gate A on phase N is driven directly or through static gates by a domino gate B on phase M, and $N - M > 1$, then phase skipping occurs. The data from gate B may go to precharge before gate A evaluates and can hence be lost. To avoid this situation we must ensure that every domino data goes through all the phases. In the case described above, two solutions are available to fix the phase skipping. If gate B has enough slack on its output it may be assigned to a later phase. Alternately, if there are static gates on the path between B and A, they may be reverted to domino gates to complete the phase sequence. If neither of the above fixes the problem, domino buffers will have to be inserted in order to complete the phase sequence.

As a consequence, we also ensure that no gate can receive a precharge (falling transition) from more than two different phases. If dom_skip_if_static_driver is set to true, the script will tolerate a single phase skipping (from N to $N + 2$) if it occurs through a static gate and if this static gate has only domino signals clocked by the same clock phase driving it.

If the output of a domino register drives a branch that is purely static, and is meeting timing requirement, it is usually wise to keep that branch static. In order to do so we must make the input to this branch static too. Therefore, the domino register is replaced by an equivalent register with one more output (Q), which is the static version of QT (i.e., does not fall when the clock goes low). This feature can be turned on/off with the dom_use_mixed_reg variable. Since converting a domino register to a static register will perturb the design, mixed registers will only be used when sufficient timing slack is available. This is specified with the variable dom_interface_delay, which has a default value of 250 ps.

In addition, a number of other variables control the phase-assignment process. These variables and their appropriate usage is described below:

dom_skip_limit: If set to true, no phase skipping will be permitted. A domino buffer will instead be inserted.

dom_skip_if_slack: Allows phase skipping when connecting to a block endpoint if the slack is less than the specified value. For example, assume U1 is a domino gate assigned to φ3 which drives a domino flip-flop, REG. We have here a phase-skipping situation which can be fixed either by shifting U1 to phase φ4 or inserting a domino buffer. If shifting the phase of U1 causes the slack on REG/D to become negative, and if the current slack on REG/D is greater than that defined by dom_skip_if_slack, then the phase skipping will be accepted.

dom_slack_tolerance: Phase assignment will abort if the WNS goes below the specified value. Note that the WNS may change after the static inputs have been delayed.

dom_preconnect: Reuse previous phase-assignment value for inserted domino buffers. Assume buf_netA_3 has been created during a previous iteration. It will automatically be assigned to φ3 independently from the arrival time on its input. When the phase assignment is redone with the new netlist, it is likely that the initial phase computation based on the arrival time on the input will assign this buffer to the same phase as its driver and hence the phase skipping will reoccur, leading to an additional buffer to be inserted. To prevent this happening, the buffer will be identified as a "phase-skipping" element and be assigned to the phase it was intended for.

4.5.11 Phase assignment: detailed description

The phase assignment requires two tools: PrimeTime™ to analyze/phase the design and Physical Compiler™ to perform the ECOs required by the phase assignment. It is recommended that both tools start concurrently in two terminals. The starting point for the phase assignment is the database produced by the placement step. The flow involves the steps shown in Figure 4.10.

Phase assignment starts by checking that the domino and static input and output constraints are as expected and that excessive negative slack is not present on any timing arc. The domino clocks are then created, and the database loaded into PrimeTime™ along with the user-defined constraints. The input and output delays are then checked with respect to the port types (static or dynamic), and the worst negative slack is checked to see if it is greater than the specified dom_slack_tolerance value. If any of these checks fail, the script aborts. The next step involves delaying all domino cells driven by static input ports. The program will abort if this causes the worst negative slack to exceed the specified slack tolerance.

The detailed phase-assignment script now assigns phases to the domino cells based on the latest data arrival at the cell inputs. If the phase assignment has been running incrementally, then the phase assigned in the previous iteration is also kept. Latches are also phased according to the domino precharge they receive. The phase-assignment

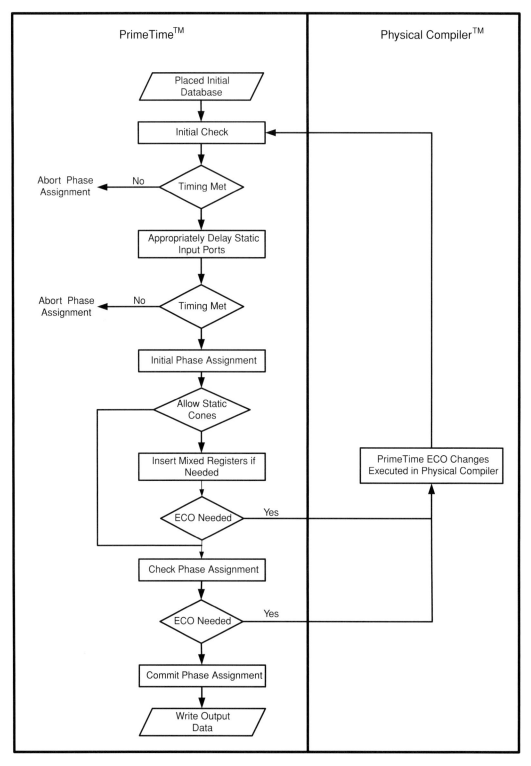

Figure 4.10. Flow chart showing the steps in the phase-assignment procedure.

program next checks the fan-out cones of domino registers to see if some branches can be kept static. Those fan-outs are kept static by having the driving domino register replaced by a mixed one. If any replacements need to be performed, an engineering change order (ECO) file is produced. Physical Compiler™ is invoked in the background to perform the ECO changes. The script then resumes at the beginning. The phase sequencing will be checked subsequently. If some static cells need to be reverted to domino cells, or if some buffers have to be inserted, again an ECO file is produced.

Once the domino cells are assigned to their final phases, any static outputs assigned to latches clocked by the last phase may be removed if dom_simplify_latched_outputs is set to true. Before exiting the phase-assignment program a last set of sanity checks is run (mostly to verify again the phase sequencing). The database results are then written through Physical Compiler™. A Verilog netlist file, PDEF floorplan file, SDC constraints file, and an Astro™ attx format clock attributes file are generated.

All Physical Compiler™ ECO changes requested by the phase-assignment script use the "legalize_placement" command. This ensures that no timing optimization is performed which would impact the timing assumptions used by the phase-assignment process. While this may make the timing in the design slightly worse, this loss is recovered in subsequent physical optimization steps.

4.5.12 Formal verification

The front end part of the flow is now complete and we are ready to proceed to routing the design. At this point it is recommended to check the phased netlist versus the initial static netlist. For the domino logic flow, this is done using Synopsys's Formality™. STMicroelectronics has a handoff script which is used to ensure that all the views needed by the back end tools are properly provided. This handoff script is incorporated into the domino synthesis flow to check possible problems.

4.5.13 Back end flow execution

The back end flow is based on an existing Astro™ flow called the AvantiKit. Upon starting back end design, a manual verification that the design only uses synchronous clear should be done. The front end part of the flow automatically generates a clock attribute file, so there is no need to run any clock tree edit tools. The clock attributes file sorts the domino clocks according to their fan-out (with the smallest fan-out first). The register clock is placed last. The target insertion delay and skew are set to 0. This setting is usually the one that produces the best results with Astro™ CTS. The specified maximum transition comes from the dom_clock_max_transition setting. The script does not add the High Fan-out Net (HFN) in the clock attribute file. The HFN synthesis flow is up to the user.

The domino cell timing models are supported in some Astro™ versions. Once that version of Astro™ is invoked, the place and route flow is similar to a standard Astro™ flow. The first step is to import a gate-level Verilog netlist and timing constraint. The phase PDEF floorplan file is then read. This file provides the location of the input and output ports on the block boundary. The location of the domino clock phase input ports should be verified. Floorplanning adds power rails and antenna diode blockage locations around the block boundary.

The clock tree synthesis is run next. Each domino phase is successively synthesized. After synthesis, one or two runs of clock tree optimization (CTO) improves the intraclock skew. The interclock skew balancing command will then further improve the interclock skew. Hopefully, the clock tree synthesis will lead to acceptable results, although some buffers may need to be sized manually for best interclock skew.

The routing is done with length-based static buffer insertion. This ensures that after a specified length, all wires will be buffered. Ensuring that long wires are properly driven reduces the likelihood of crosstalk failure on those wires. As in standard routing flows, the clocks are routed first for the domino design, followed by the other nets. To maximize yield, double vias are inserted wherever possible. The router will attempt to analyze crosstalk violations in the design. Subsequent optimization steps allow the crosstalk violations to be corrected. The routing will also fix antenna violations and insert filler cells in empty cell locations. Timing optimization occurs during routing, to ensure that all timing constraints are satisfied.

4.5.14 Minimum pulse width overlap check

As described in Chapter 3, for a domino cell to evaluate properly the input and clock must overlap sufficiently so that the output pulse will have enough energy to trigger the next gate. Since this check tends to fail for some cells the first time it is run, it is best to be run when the Astro^{TM} session is still active and changes can easily be made. To run it a Verilog netlist, a constraints file, and a parasitics file are required (DSPF format). The MPWHO check is performed in PrimeTime^{TM} using a script called dom_check_mpw. The minimum pulse width overlap check can be run on the different process, voltage, and temperature corners. This check can be modified with two global variables, a global scaling factor and a report flagging variable. The global scaling factor has a default value of 1. Using values greater than 1.0 ensures that the measurement is more conservative. The report flagging variable allows all the area computations to be reported, and not only the failing ones. This is useful to understand if there are many points close to failing.

The dom_check_mpw script produces a violation file report and a list of all pins that should be slowed down to fix violations. Static buffers are inserted in Astro^{TM} on all nodes that need to be delayed. If routing produces a design with many failing nodes, especially those in which no overlap is present between the inputs in some domino cells, one pass of buffer insertion may not be enough. After each iteration of inserting static buffers the check needs to be rerun until all nodes pass. To avoid a respin after signoff, some margin can be obtained by setting the global scaling factor to a value slightly greater than 1.

4.5.15 Crosstalk check

Voltage bumps can cause domino cells to discharge, leading to possible functional errors in the design. To prevent failures, each input pin has a maximum voltage bump requirement that must be met. In order to avoid being excessively pessimistic when the induced voltage bumps are calculated, timing windows-based analysis is needed. Timing windows consider the specified periods of time in which different aggressors can act on a victim node. This enables a far more accurate, and generally less severe, maximum voltage bump to be calculated for each victim node. A PrimeTime™ script, dom_check_xtalk, is used to calculate the maximum bump heights with timing windows. The script needs a file listing the maximum acceptable input voltage bump on all domino cells and an Astro™ crosstalk report listing all possible bump height violations.

Two additional variables can be set with this file: dom_req_derate, which allows for the bump requirement to be derated, and dom_win_stretch, which allows the timing windows to be stretched to increase the timing overlap between victims and aggressors. Crosstalk violations can be corrected in Astro™ through an iterative process by rerouting failing nets.

4.5.16 Signoff verification

Final signoff for the domino modules performs the standard post layout design checks. Design rule check (DRC), layout versus schematic (LVS), and antenna violations are checked with Mentor Graphics Calibre. The parasitics for the routed design are extracted via Synopsys's StarRCXT. Timing and formal analysis are performed to ensure that the design flow did not alter functionality. It is also highly recommended that back annotated dynamic simulations be performed to further verify functionality. The domino-specific checks: minimum overlap and crosstalk are again run with the signoff netlist and parasitics.

Block packaging requires that a timing model for the domino module be provided. Unfortunately, the current version of PrimeTime™ is not able to generate any correct model (.lib or ETM) for domino blocks. Another PrimeTime™ script has been developed to create a black box model for the domino design. This script will produce the minimum or maximum delay model. Each timing model contains all the timing arcs required for the module to be correctly timed in a pure static environment. It also specifies acceptable clock relationships.

4.5.17 Final comments

The primary challenge encountered in the development of the domino synthesis flow has been to get the tools to understand the timing models needed for domino logic. While the timing analyzers have generally been able to handle the complexity of the timing model, it has been more difficult for the implementation tools to support the full timing model. This has been a problem particularly for the physical design tools, since after phase assignment the full domino timing model is used. Using the pseudo-static timing model for the synthesis and initial placement has meant that those tools do not need to support the full timing model. One way to understand the difference between static and domino logic is that static logic is level-based, i.e., inputs can go high or low, with the final result reaching certain steady-state levels. Domino logic, on the other hand, is pulse-based, with an evaluate phase rising transition followed by a precharge phase falling transition. The current timing analysis tools are designed for static cells, and hence are not naturally suited to measuring the quality of pulses. For some of the more complex checks, specifically those that involve the overlap of domino phases, even the timing analyzers cannot support these features natively. Extensive scripting using the application processing interface (API) for the timing analyzer has to be used to support these features. Despite these challenges, a workable domino logic synthesis flow has been developed. If ASIC designers start using domino synthesis, the EDA companies are expected to provide more support for domino-compatible timing models. One other point to mention is that while the flow described used particular EDA tools, extensive experiments were conducted using a number of different EDA tools. Many of the steps described can easily be ported to other EDA tools. The final flow, as described, is derived from a static ASIC flow, and hence the tool choice is based on tools already selected.

4.6 Schematic capture of domino designs

It may seem somewhat odd to describe how to use schematic capture for domino cells at the end of a chapter discussing how domino logic should be synthesized. We do not, however, believe it represents a contradiction, and start our defense with a joke.

The joke involves an engineer, a physicist, and a mathematician who are staying in a hotel. A small fire breaks out in each room, with the engineer using water to douse the fire in his room. The physicist spends much time thinking about an optimal solution before using the fire extinguisher. The mathematician wakes up and sees the fire. He notes that the room has a fire extinguisher and goes back to sleep without actually bothering to extinguish it, reassured "that a solution exists". This joke is one of a series told by engineers, portraying mathematicians and physicists (preferably theoretical physicists) as being irrelevant theoreticians. Of course, as part of their response, physicists and mathematicians have their own jokes about engineering, most of which focus on what can best be described as the limited theoretical rigor of engineering compared with their own fields.

The point of the joke, of course, is that engineering is a practical field. While domino synthesis may be useful for very large designs, if a small module has to be implemented with domino logic, schematic capture may be the best choice, especially if the designers do not have access to a pre-existing domino synthesis flow. When schematic capture is to be used for domino designs, the final netlist has the same requirement as a synthesized domino logic netlist: the design should be unated and the domino cells must be clocked to ensure that evaluate or precharge failures do not occur. These two steps are basically the unate and phase-assignment process in the domino synthesis script. The optimizations possible with a schematic captured design depend on the clocking scheme. If a single clock source is provided to the module, it is relatively easy to generate the clock and its inverse. Depending on clock speed, other phases can be generated with varying degrees of ease.

For a schematic captured domino design, the designer should keep in mind all the ways we have listed that domino design can fail, and check to ensure that the design is robust under different process and environmental conditions. Several points arise as to how the design should be laid out and validated. One of the main advantages of custom, or structured custom, designs is that the wiring can generally be minimized. If the logic is implemented as standard cells (custom or from a library), an ASIC place and route tool can be used to route the wires. For larger designs, that is a major advantage. The designer is, however, warned to double-check that the wiring path and layers used by the router are acceptable.

One of the challenges in custom design is validating the functionality and timing of the logic. For relatively small designs, extensive dynamic simulation is probably the easiest path to validate both aspects. For larger domino designs, some amount of dynamic simulation is also useful as it serves as a sanity check for the behavior of the circuit. Design validation for larger custom blocks is generally much easier with some models for the standard cells. The requirements of the models depend on the size of the design and also the complexity of the functionality entailed. For example, a domino circuit implementing an incrementor or some clock divide logic behaves much more repetitively in terms of the output logic states than say a small multiplier. Verilog is good for checking functionality, with timing models (if not complete at least with some key features) being very useful. A final point to note with any design, whether it is custom or a pure ASIC one, is to make sure that the design starts correctly at power up, after reset and after any other interrupt condition. This is particularly important for domino design, as specific relationships may need to exist between clock phases and data which have to be carefully checked.

In the next chapter we describe the results obtained by implementing domino logic with a number of different ASIC-compatible design flows. These include a schematic capture-based approach, as well as a complete ASIC-type synthesis flow.

References

1. D. Harris, *Skew-Tolerant Circuit Design*, Morgan Kaufmann Publishers, San Francisco, CA, 2001.
2. R. Puri, A. Bjorksten and T. E. Rosser, Logic optimization by output phase assignment in dynamic logic synthesis, IBM Research Report, RC 20533 Computer Science/Mathematics, August 1996.
3. M. Zhao and S. Sapatnekar, Dual-monotonic domino gate mapping and optimal output phase assignment of domino logic, IEEE International Symposium on Circuits and Systems, 2000.

5 Circuits designed with domino logic in an ASIC flow

5.1 Introduction

Previous chapters in this book have been devoted to the design of domino logic standard cells and methods to synthesize logic using them. In this chapter we describe some example circuits implemented using different automated domino logic design flows. Since the primary benchmark for synthesizable domino logic is against synthesizable static logic, comparisons are provided between the two. Silicon-measured data is also provided wherever it is available.

5.2 Domino integer execution unit

A typical application for high-speed logic is in the execution units of microprocessors. Execution units are the main arithmetic modules in processors, performing integer or floating point arithmetic. In order to understand the speed advantages possible with domino logic, we decided to build a simple integer execution unit. The block has an adder, a shifter, a multiplier, and a bit operations unit. Memory modules interact closely with execution units, to provide data and instructions. For this design two 32-entry, 32-bit wide register files are used in each execution unit. One register file supplies the 32-bit wide data operands that are applied to the datapath modules and stores the result. The other register keeps a simple set of instructions. These instructions allow the data operations to start and stop. They also determine the operations to be performed and the data memory locations to be accessed.

A schematic representation of the execution unit data flow is shown in Figure 5.1. Operation starts via instructions sent from the instruction register file. Each arithmetic function receives operands from the data register file. The shifter and bit operations unit also receive control signals from the instruction register file. These controls determine if a left or right shift is needed in the shifter, or if a bitwise AND, OR, XOR, or inversion function is to be performed in the bit operations module. The adder and multiplier operate on 2s complement signed arithmetic operations. Since the output of the 32-bit adder and multiplier exceeds 32 bits, the output is truncated to the least significant 32 bits.

One of the main purposes for designing an execution unit test chip was to determine the speed advantages of domino. To do this in silicon, a static logic execution unit was also placed on the same die. The design of the execution unit predated the development of

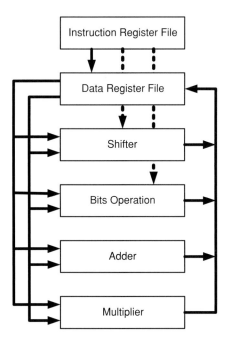

Figure 5.1. A simple microprocessor execution unit.

the full synthesized domino logic flow. For this reason the gate-level domino netlist was produced using schematic capture, with automated ASIC tools used for physical design. The static logic version of the execution unit was implemented with logic synthesis. The test chip was built using a 0.18 μm CMOS process.

The execution unit was our first domino logic design, so no standard cell library of domino logic cells was available. This meant that a set of specific domino cells had to be designed and laid out. This approach for using domino logic, where some domino logic standard cells are used to implement a logic module in a larger ASIC, may be useful for some designers. This is especially true if they do not have the resources or time needed to develop a complete domino synthesis environment.

5.2.1 Dual-output domino logic

In addition to the single-output domino cells described in Chapter 3, the domino execution unit used dual-rail domino cells. A dual-rail domino AND cell is shown in Figure 5.2. In dual-output domino cells, both outputs are low when the cell is in the precharge. During evaluation, one of the outputs must go high (if the logic is working properly), with the other output remaining low. Generally, both the true and false outputs are provided to the next cell, which again provides dual outputs. This makes the schematic capture process for tracing wires across logic cells easier when duplication is present. In dual-rail cells, the pull-down clock transistor (called the footer) can be shared. For non-critical cells, this saves transistor area compared with two separate dual domino cells. This advantage is

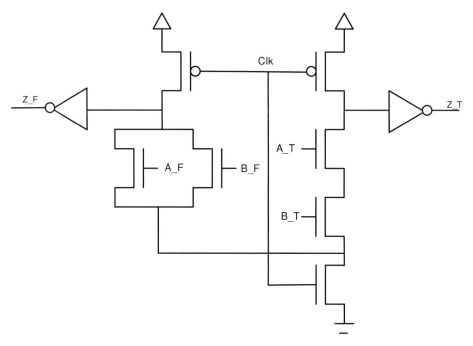

Figure 5.2. A dual-output domino AND gate.

diminished if the domino cell needs to be very fast, since the size of the footer transistor has to grow.

The synthesizable domino logic solution did not use dual-rail domino cells in the flow, since it would lead to full duplication of the logic, even when trapped inverters could be removed by the bubble-pushing process. In addition, the size of the library tends to increase since optimum cell design requires dual cells to be available with different sizes for each output driver. Let me explain this point with a simple example. If a dual-rail cell has two outputs, one with a number of NMOS transistors in series in its pull-down path, the inverse path will have pull-down transistors arranged in parallel. The output driven by parallel NMOS pull-down transistors will be faster than the stack with series NMOS transistors. In order to balance delays from inputs to outputs in dual domino cells, it is best to independently size transistors for the two outputs. If a base domino cell has four drive strengths, then an optimized dual-rail domino cell could require any of the four drives on either the true or false paths. This means that the dual-rail output would need 16 drive strengths. While one can say that having four base drive strengths for the dual-rail cells will ensure that no speed degradation occurs, this will lead to an over-optimized design, with greater area and power.

Despite this disadvantage, dual-rail domino logic does allow for a relatively straightforward unate domino implementation. The penalty for duplication may also not be severe for datapath implementations, since these circuits generally require full, or close to full, duplication. Under such circumstances, a dual-rail library can be an effective solution. This was the case with the execution unit, a relatively small module, which only needed 13 different domino cells.

5.2.2 Schematic capture and library specification

The fast adder design was based on the Kogge–Stone algorithm. The logic partitioning was done simultaneously with standard cell schematic development. Expected parasitic values were annotated in the schematics to ensure that the difference between the schematic and the final physical layout was minimized. The Kogge–Stone adder was also used to implement the fast adder needed in the carry propagate stage of the multiplier. The multiplier design was based on using a domino carry save 4-to-2 compressor [1]. No other compressor cell was used. Based on the requirements of the design, a set of domino logic cells was developed. All domino cells were layout-compatible with a static logic cell in the same technology. The following list of single-rail and dual-rail domino cells was designed for the domino execution unit:

Single-rail domino buffer: Two drive strengths for the single-output domino buffer were used. A standard domino buffer (equivalent to a 4× static library cell in terms of total transistor width) and a high drive strength buffer (equivalent to a 16× static logic buffer in the static library). The larger cell was needed for heavily loaded nets, especially in the multiplier.

Single-rail two-input domino AND cell: The bit operations unit needed a two-input standard drive domino AND cell.

Single-rail two-input domino OR cell: This cell was also needed in the bit operations module.

Single-rail two-input unencoded domino MUX cell: The functionality of the cell is: $Z = S0 \times D0 + S1 \times D1$, where D0 and D1 are the data inputs to the cell and S0 and S1 are the multiplexer select signals. Although S0 and S1 are inverses of each other, they are provided to the cell to avoid using an explicit inverter. The multiplexer cell was used in the output selection logic.

Single-rail three-input unencoded domino MUX cell: This cell is similar to the two-input multiplexers, but with three data and three select inputs.

Dual-rail two-input AND cell: The presence of non-removable trapped inverters in some modules meant that a dual-output AND was needed. The evaluate cycle true and false output values are: $Z\_T = A\_T \times B\_T$, and $Z\_F = A\_F + B\_F$. This cell is shown in Figure 5.2.

Dual-rail three-input AND cell: A three-input dual output AND cell.

Dual-rail CSA42: Compressor cells used in the design of the carry save portion of the multiplier. The CSA42 is a carry save cell that actually has five inputs and three outputs, but that reduces four carry save inputs to two outputs [1].

Dual-rail GEN0 cell: A number of dual-rail generate cells are used in the Kogge–Stone carry propagate adder. The outputs of the GEN0 are: $Z\_T = A1\_T \times B1\_T + A0\_T \times B0\_T \times (A1\_T + B1\_T)$ and $Z\_F = (A1\_F + B1\_F) \times (A0\_F \times B0\_F + A1\_F \times B1\_F)$. This cell is large, as it has eight inputs and two outputs.

Dual-rail GEN3 cell: A dual-rail cell that implements the true function $Z = G1 + P1 \times G0$. The G and P inputs refer to generate and propagate inputs. In adders, a generate signal always causes the carry-out to be high. A propagate term creates a high carry-out signal only if a carry-in to the cell is also high.

Dual-rail GEN3 cell: Another generate cell used in the fast adder. The cell's true functionality is $Z = G2 + P2 \times (G1 + G0 \times P1)$.

Dual-rail PROP0 cell: An eight-input propagate logic cell. The function of the cell is: $Z\_T = (A0\_T + B0\_T) \times (A1\_T + B1\_T)$ and $Z\_F = (A0\_F \times B0\_F) + (A1\_F \times B1\_F)$. The PROP0 cell received the primary inputs to the fast adder. The propagate signals generated were valid for two adjacent pairs of input signal, e.g., inputs 0 and 1, to the adder.

Dual-rail two-input XOR: The final output of the carry propagate adder is a two-input XOR cell. This cell was also used in the bit operations module.

Dual-rail two-input MUX: This cell is the primary cell in the shifter.

While there are 13 different domino cell functions listed, the dual-rail two-input MUX cell, the two-input XOR, and the PROP0 cell have the same transistor structure. These cells, hence, shared the same layout with only input pin names being changed.

5.2.3 Delay, power, and crosstalk analysis

Detailed transistor sizing for the domino logic cells was done by placing three domino cells in series. The first cell provides a realistic transistor drive to the second cell, whose delay is being measured. Digital logic must always be designed with realistic input drive strengths and input transition times. This is important, as Spice-like simulation tools have infinite drive capacity, which masks the extra loading delay caused by using larger input transistors. The second cell is the device under test (DUT). The third cell provides realistic output loads. We assumed a fan-out loading of two cells at the output of the device under test. In addition to the load of the active elements, the load of the wires in the design needs to be considered. For small process geometries this interconnect load is a major portion of total output load. The capacitance for every output wire was approximated by multiplying the average length of output wires times the average capacitance per unit length. As a conservative estimate we used wire lengths from the multiplier (which have the longest wire lengths) for our purposes. The average capacitance was found to be 7.8 fF (for the 0.18 μm process we used an average capacitance of 0.22 fF/μm for each wire). Based on these assumptions, each domino cell was sized. During the design process it was possible to get an approximate delay for every cell. This could then be used to estimate module delays. Design changes could then be made as needed.

The design of the domino cells preceded the more sophisticated crosstalk modeling strategy described in Chapter 3. Still, we needed to make sure that no crosstalk-induced failures occurred in the design. A conservative guideline for ensuring crosstalk tolerance was used in order to guarantee functionality. We were also constrained at the time as the router used was not crosstalk-aware. The primary attribute we focused on for determining the crosstalk tolerance of the design was the NMOS width of the domino cell inverter. When a rising spike is coupled onto a line driving a cell, the ability of the victim line to remove the coupled charge depends on the strength of its driving NMOS transistor. The coupling noise simulations assumed a 500 μm wire, 80% of whose total capacitance is coupled to two aggressors. Furthermore, these aggressors are driven in

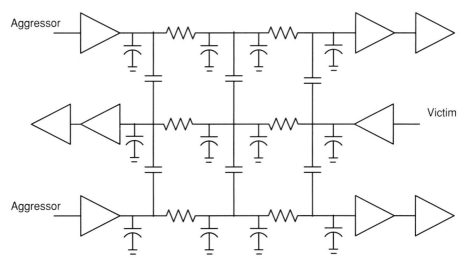

Figure 5.3. Crosstalk noise simulation for domino cells in the designed 0.18 µm execution unit.

the opposite direction from the signal. This represents a worst-case condition, since the distant weakly driven segment of the victim line is coupled to the strong aggressor. Figure 5.3 shows the simulation setup. To mimic the RC effect of wires, the wires were modeled as a distributed, coupled RC network. The victim has its input tied to GND, with the aggressor set to Vdd. The receiver cell was the small domino buffer, with minimum loading at its output (higher capacitance helps filter input voltage spikes). These conditions ensured severe noise coupling conditions, much worse than anything likely to be encountered in the actual circuit. It did, however, ensure a very rugged design. For the domino cell designed for the execution unit, coupling noise-induced failure was said to occur whenever the output voltage of the receiver cell exceeded 36 mV, or 2% of the nominal Vdd supply. The problem formulation for these cells was simplified by the presence of a single domino logic driver size, except for the large buffer driver, which meant that we did not have to consider many different drive sizes for aggressors and victim lines.

5.2.4 Transistor sizing guidelines

In Chapter 3 we discussed the design of domino standard cells primarily in terms of the characterization requirements that a domino logic cell included in a domino synthesis system must adhere to. For the execution unit, such a detailed characterization flow had not yet been developed. The primary requirement for the domino cells was to be fast and stable. We shall next describe how transistor sizing was done for the cells in the execution unit. Unlike in Chapter 3, this section provides a more detailed description of the cell design process.

In Figure 5.4 a general domino cell is shown. The transistor sizes in the design, as shown in Figure 5.4, correspond to five transistor widths: W_p, W_n, W_{pi}, W_{ni}, and W_{pk} that can be specified for each cell. These five transistor sizes correspond to the

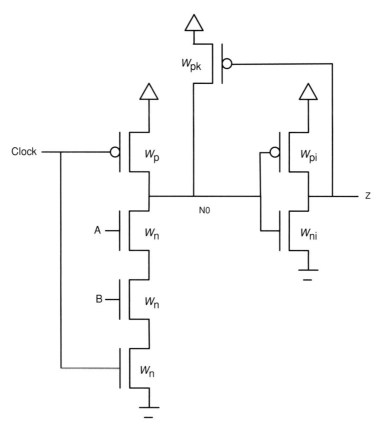

Figure 5.4. Domino logic AND with a weak feedback keeper.

PMOS precharge pull-up transistor (W_p), the NMOS evaluate transistor size (W_n), the output inverter PMOS transistor (W_{pi}), the output inverter NMOS transistor (W_{ni}), and the weak feedback transistor (W_{pk}). Starting with some published guidelines [2], we developed some insight into the ratio of transistor widths that provided a good tradeoff between performance and stability. The reader is reminded that these ratios may vary with different CMOS processes and simulation corners. As always, care must be exercised when applying general guidelines.

The sizing of the dynamic gate of a domino cell (W_p to W_n) strongly determines the ratio between the evaluate and precharge speed of the domino cell. David Harris has recommended using an effective W_p to W_n ratio for the dynamic gate that is less than the mobility ratio by a factor of two [2]. The tradeoff in this ratio is between preferring evaluate delays (a lower ratio) and favoring precharge delays (using a high ratio). For the dynamic execution unit, the ratio of W_p to W_n width used was 1.5 to 2. This was based on using a precharge delay of 250 ps, which in retrospect was too high for the process used. This illustrates how an accurate estimate of precharge delay must be used when a domino library is designed. Failing to do so will lead the library precharge or evaluate delay to be the critical path. A difficulty we encountered in using specific ratios is that depending on the cell topology, this did not correspond to the same worst-case precharge delay. The cells were, hence, optimized to make sure they all had approximately the same precharge delays.

The next transistor ratio considered was for the output inverter: W_{pi} and W_{ni}. In order to determine the sizes for W_{pi} and W_{ni} a number of factors needed to be considered. The domino cells must be able to drive their expected loads with reasonable delays and transition times. Also, the ratio of W_{pi} to W_{ni} determines the noise susceptibility of the cell. A higher ratio improved the evaluate delay of the cell, while a smaller ratio improved the crosstalk tolerance of the cell. For the domino execution unit the ratio chosen was 3.

The sizing constraint for the weak feedback keeper (W_{pk}) was based on the need for the keeper to counteract leakage when the output remains low during evaluation. Without the keeper the internal node would be floating during this period, leading to a possible failure if leakage current caused the node to go low and the output high. Even without such an extreme situation, the floating internal node leads to the output NMOS transistor in the inverter being more weakly driven and less capable of fighting crosstalk-induced voltage bumps on its output. The feedback inverter also helps the domino cell mitigate the effects of charge sharing. The sizing needs for this cell were determined by the coupling noise simulation shown in Figure 5.3. The tradeoff in the weak feedback transistor size is between the stability of the domino cells (which requires a larger W_{pk}) versus a faster evaluate delay (which prefers a smaller W_{pk}). Harris has suggested a keeper size of 1/4th to 1/10th of the effective strength of the pull-down stack (W_n). For the CMOS 0.18 μm domino cells in the execution unit, we used minimum-sized transistors as this represented a good tradeoff between performance, stability, and ease of layout.

In addition to the transistor sizes shown in Figure 5.4, the domino cells for the execution unit had internal precharge transistors to counteract charge sharing. These transistor widths were always minimum-sized (larger values increased delays). Internal precharge pull-up transistors were used whenever four or more NMOS transistors were in series (including the footed evaluation transistor driven by the clock). For best performance the precharge transistor was placed in the middle of the stack. Internal precharge transistors were also used for nodes that were connected to four or more NMOS transistor drains. The exception to this rule was if the node was connected to the footer device, since it led to large shortcircuit currents at the beginning of precharge. In subsequent designs the use of internal precharge nodes was discarded. This was due to the layout overhead in including them and the fact that data could arrive long after the clock was enabled, leading to the internal voltage largely being discharged through leakage. Leakage in CMOS processes has become considerably worse after the 0.18 μm process. The technique may still be useful when precise cell timing is known, or in older CMOS processes.

5.2.5 Design of the execution unit

As mentioned, the execution unit was designed using schematic capture for the domino execution unit and synthesis for the static module. Physical design for both modules was done using standard ASIC design tools. At the implementation level there were only two differences between the static and domino units. Firstly, we wanted to ensure that unnecessary switching power was not consumed in modules whose operation was not selected. This required that those modules be shut down. For the domino block it is possible to stop all switching in a module by shutting down the clock to the module. This technique cannot be used in the static unit because the data is provided to the functional modules directly from the register file. To guarantee no switching inside the functional modules, their input data were ANDed with an enable signal. The enable signal used is derived from the instruction being selected for a particular operation. The second difference was testing methodology. The block functionality was determined with functional testing. For failure analysis a scan chain was included in the static execution unit. The scan chain could not be included in the domino unit because no explicit flip-flops were used in the design. The presence of the scan chain in the static implementation makes the comparison a little unbalanced, favoring the domino modules.

The four clock phases needed in the test chip were generated from the global clock by using a clock-generation block. The clock generator uses programmable delay elements and inverting phases to generate four clock phases from a single source. It could support clock frequencies from 300 MHz to 1.2 GHz. The clock phase had a duty cycle variation across process, voltage, and temperature from 50% to 65% (at 1 GHz). This was considered acceptable for the design.

While schematic capture was used to implement the domino modules, the design then followed closely a standard ASIC flow. The cells were placed using Synopsys's Physical Compiler^{TM}. Since Physical Compiler^{TM} is a timing-driven optimization tool, simplified timing models for the domino cells were developed. Cadence's Silicon Ensemble was used to subsequently route the design. Domino cells in the same clock phase were grouped together. Figure 5.5 shows a layout plot of the chip with the two execution units.

In order to test the functionality of the execution units, the data and instruction registers were loaded via a slow external serial interface. The execution unit was then operated using the high-frequency PLL clock. Finally, the results were read out serially. This procedure was adopted as it was not possible to verify high-speed chip functionality through the I/O and the tester.

5.2.6 Silicon results

Silicon tests on the static and domino execution unit showed that both designs were functional. The maximum operating speed of the domino and static adders at 1.8 V was 1360 MHz and 482 MHz, respectively. This demonstrated a significant speedup in the domino design. The domino bits operation unit and shifter also operated at the adder speed, indicating that the maximum operating speed was limited by the register file read operation (a 32-bit adder incurs at least five gate delays versus a single gate delay for a bit operation). Under worst-case environmental conditions (Vdd at 1.65 V and 125°C), the domino design was operational at 960 MHz and the static design at 390 MHz.

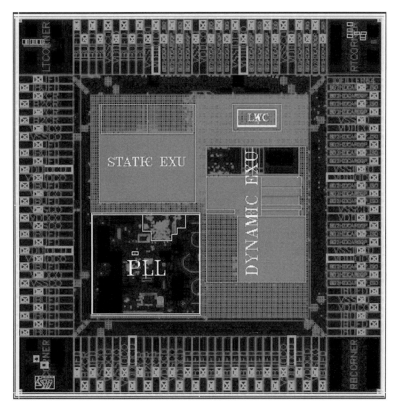

Figure 5.5. Layout plot of the execution unit showing the static and domino execution modules.

The pipelined domino multiplier operated at 950 MHz while the static multiplier ran at 425 MHz.

In addition to the speedup achieved with the domino design, we were happy with the stability of the solution. The domino design was operational at 0.8 V and at a low speed of 500 kHz. One of the experiments run was to test the operating voltage required by the domino and static adders to operate at 320 MHz. For the static design the supply voltage needed was 1.36 V while the domino design only needed a 1.02 V supply. Figure 5.6 shows a plot of the maximum operating frequency for the 32-bit adder for all the domino designs across an 8-inch wafer. The plot shows that while only one other die achieves the exact same high frequency, almost every other die can operate faster than 1.2 GHz. The two die with no numbers associated with them were non-functional. The distribution of frequencies across the wafer mirrored results seen in static implementations. This was reassuring. Figure 5.7 shows an execution unit die on a tester.

Power measurements on the test chip showed the unbelievable result that the static core was consuming more power than the domino design (74 mA versus 46 mA). If something is too good to be true, in engineering at least, generally there is a problem. In this case the issue was traced to be the use of a grossly exaggerated clock tree network for the static design. This was due to incorrect operation of the clock tree synthesis

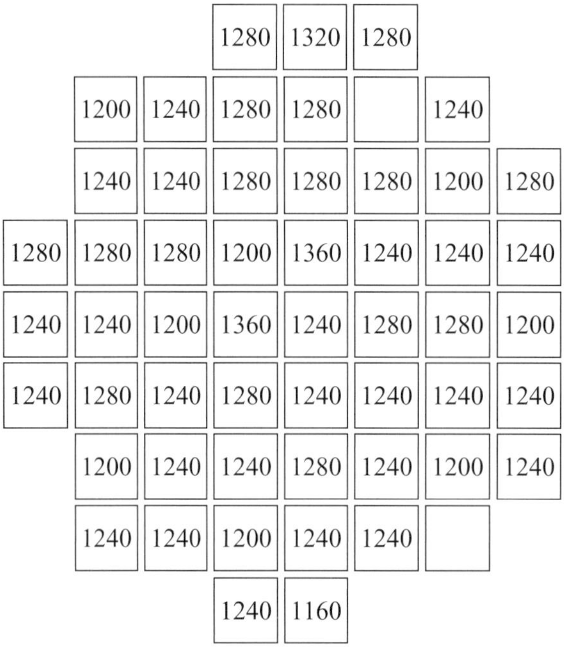

118 High Performance ASIC Design

			1280	1320	1280		
	1200	1240	1280	1280		1240	
	1240	1240	1280	1280	1280	1200	1280
1280	1280	1280	1200	1360	1240	1240	1240
1240	1240	1200	1360	1240	1280	1280	1200
1240	1280	1240	1280	1240	1240	1240	1240
	1200	1240	1240	1280	1240	1200	1240
	1240	1240	1200	1240	1240		
			1240	1160			

Figure 5.6. Maximum operating frequency distribution across a silicon die.

Figure 5.7. The execution unit bare die on a tester.

tool, which resulted in an unnecessarily deep and heavily buffered clock tree network. A careful study of the actual power dissipation indicated that the domino design would consume twice the power of the static design.

The test chip results showed that a domino design could achieve more than twice the operating frequency of an equivalent static synthesized design. While the flow had involved schematic capture, automatic tools were used in the design and verification flow, including a simple version of the standard cell models described in Chapter 3. After tape-out we ran experiments to see how the speed of the design would change by allowing the tool to place the cell instances and using an automatic clock tree synthesis flow. The results of these studies were encouraging and led us to start exploring the use of a fully automated domino logic flow.

5.3 A synthesized domino logic DSP core

Following the design of the domino logic execution unit we started working on the first version of the synthesizable domino logic flow. An application of the flow was found in the design of a specialized multimedia digital signal processor (DSP) core. A few modules in the core needed to be sped up and synthesizable domino logic looked useful. The DSP core is used in consumer applications, which are notoriously price-sensitive. The extra area needed to unate the design for domino logic would appear to make this unacceptable. This was not, however, the case due to a number of reasons. Firstly, the total number of modules that needed to be implemented in domino logic was small, meaning that the total area penalty in using domino logic was also small. Secondly, a failure to meet performance using static logic could have required a microarchitectural modification to the design. This would have entailed significant design, verification, and software modifications, the cost of which would be far higher than the extra area penalty. It had also been suggested that a future domino implementation could satisfy the computation requirements that currently needed two static cores fulfilled. This implied a possible reduction in total chip area by using domino logic!

Three modules in the DSP core were implemented using domino logic. The design was done in a 0.13 µm CMOS process. The three modules implemented with domino logic used 24,000, 9000, and 13,000 standard cells, respectively. For static and domino synthesis the same RTL and basic constraints were used, with only a few domino-specific constraints being added. To complete the domino designs required runtimes that were less than 3 hours longer than for the static designs. This was considered very acceptable. The area required by the domino design was 2.06 greater than the static design after placement and 2.3 greater after routing. The domino design was found to be routable with an area utilization of greater than 80%. This was important, since domino designs have more wires than static designs, which means that they are in general more difficult to route. Using a low area utilization for domino designs would further increase the area penalty of domino logic. The average speedup achieved using domino logic was 1.45 faster. This is roughly equivalent to the speedup achieved in two process generations.

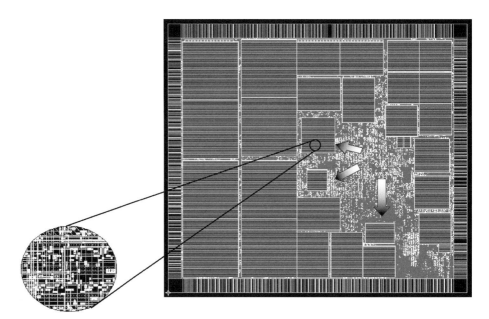

Figure 5.8. A DSP containing modules synthesized with domino logic.

Figure 5.8 shows a plot of the chip with pointers showing the three modules implemented with domino logic. A small area in one of the domino blocks is shown in greater detail. The routing for the module is seen to be typical of blocks routed with automated place and route tools. Figure 5.8 also shows that the total area consumed by domino logic is relatively small. Most of the other blocks in the DSP are memory modules. It must be remembered that this processor is itself part of a larger SOC, further diminishing the total area and power overhead in using domino logic.

While a test chip with the DSP core has been manufactured, it has not been tested. This is primarily due to it being discovered during the design process that paths through the memory modules would become critical in the new design, largely negating the utility of using domino logic in the three logic modules. The designers of the DSP did not have enough bandwidth at the time to redo the microarchitecture to fix the problem. With this limitation, only a small speed improvement was expected (approximately 10%) in using domino logic.

Despite the limited speedup we had still hoped that testing the module would demonstrate the functionality of the complicated flow. As mentioned, this did not happen due to a lack of tester time. High-speed tester time is expensive and scarce. The testers themselves cost many millions of dollars, with a full-time engineer required to operate the machine. While the reader may consider this to be a somewhat interesting side point, it does have a profound impact on high-speed digital design for ASICs. Most designers working in industry must design their chips to worst-case process, voltage, and temperature conditions. The actual silicon invariably tends to be faster. For designers and managers struggling for high speed, this discrepancy often appears solvable by using a more forgiving process corner. The difficulty with this solution is that it often requires

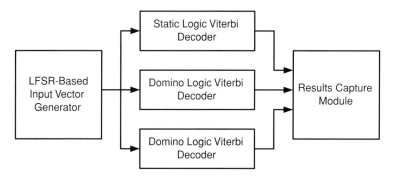

Figure 5.9. Data flow overview in the Viterbi test chip.

far greater high-speed tester time to characterize the maximum operating speed of functional parts. This process is called speed binning, which refers to the different operating speed "bins" that the parts can be placed in. The extra testing cost in speed binning is generally not acceptable, however, for any products other than microprocessors or some other high end products, since it is not possible to charge more for parts that can operate faster.

While using synthesizable domino logic in a DSP core unfortunately did not allow us to validate the results on silicon, it did inform us of two limitations in the domino flow which required flow modifications. While the speedup was close to our goal of a 1.5 times increase, the area overhead was higher than anticipated. To limit this in the future we worked on modifying the flow to include static logic for non-critical paths. This has been described in Chapter 4. Secondly, the domino logic blocks had static inputs and outputs which required to be properly accounted for. In order to do this we started to look at interface issues between static and domino logic. Having explicit flip-flops between static and domino logic reduces the interface overhead between the two, allowing domino logic to be more effectively used. This strategy was used in our next test chip, a domino logic Viterbi decoder add–compare–select (ACS) module.

5.4 A synthesizable domino logic Viterbi add–compare–select (ACS) test chip

A potentially useful application of domino logic was found in the design of a Viterbi add–compare–select (ACS) module. This block needed very high speed. It had a single cycle feedback loop that could not be further pipelined. In order to compare a static and domino synthesis, we implemented the Viterbi decoder using both flows on the same 0.13 μm CMOS chip. A schematic representation of the different operating modes of the chip is shown in Figure 5.9. The design has a static and two domino implementations of the logic. The first domino implementation used the standard domino flow, while the second was designed with a more experimental domino approach.

Viterbi decoders are used to reduce bit error rates in noisy channels. Using domino logic to achieve faster speed directly improved the data rate in the Viterbi decoder. As in the execution unit test chip, the high operating speed of the logic required on-chip test vector generation and result capture. The input data to the Viterbi decoder was generated by a 10-bit linear feedback shift register (LFSR). The LFSR block generated 1152 input signals, which were then provided to each Viterbi decoder. An FSM-based controller captured the Viterbi decoder outputs after 1024 clock cycles. The output data could then be sent off-chip using a slow serial interface. The test chip contained a PLL to generate the high-speed on-chip clocks.

The 32-state Viterbi module was specified in Verilog. The Verilog was actual production code. No modifications were made to it to support the domino logic flow. The design constraints (input and output delays, driving cell strengths, etc.) for the domino Viterbi module were based on the chip floorplan. The only domino-specific design constraints added were to specify whether each input and output was static- or domino-compatible. Since the LFSR and result capture module achieve their target speed with static logic, only the Viterbi modules were implemented using domino logic.

The 130 nm domino library used in the design was the predecessor of the 90 nm domino library described in Chapter 3, but based on the same principles. The domino logic standard cells were layout-compatible with the available static logic standard cells, i.e., being abuttable with and having the same row height. Since domino logic modules also contain static cells, the layout compatibility between the libraries ensured that an existing static logic library could be used in the design flow. This compatibility simplified chip assembly when static and domino logic modules must be placed together. There were 400 cells in the 130 nm domino logic library.

Since crosstalk-induced noise can lead to functional failures in domino logic, all of the cells in the domino library were characterized for their noise tolerance. In a synthesizable domino logic, flow timing closure becomes unpredictable if major modifications are needed to correct crosstalk-induced noise. Crosstalk-induced noise bumps can only be analyzed accurately after the design is routed, which is at the end of the design process. If correcting the crosstalk problem requires excessive buffer insertion or placement modification, the phase-assignment process will need to be repeated, which may significantly perturb the design. To reduce post-routing crosstalk noise problems, the cells were designed to be rugged. Noise characterization for the domino cells was performed assuming a number of simultaneous effects, including: multiple independent aggressors, charge sharing, noise propagation, and victim driver weakening. Crosstalk analysis and failure correction was done using Synopsys's Astro^{TM} by route modifications. Final crosstalk noise is independently checked during the signoff process. Since the domino logic library was part of an ASIC-style flow, all of the views required by ASIC tools (timing, layout, test, Verilog, VHDL, etc.) were provided.

The design flow for the domino logic Viterbi decoder started with an initial synthesis using Design Compiler^{TM} from Synopsys. Since domino is a non-inverting logic style, the design is then unated. This entails ensuring that no inverting cells are present in the design. The automation of the unate procedure was done with Tcl scripts within Design Compiler^{TM}. The physical design for the Viterbi test chip utilized Physical Compiler^{TM} for initial placement and Astro^{TM} for clock tree synthesis and routing (both from Synopsys). The only other domino-specific task that needed to be performed in the domino synthesis flow was to assign clock phases to the domino cells. This was done in an automated iterative loop with cell placement. The domino Viterbi decoder used a

Logic Synthesis	Design Compiler (Synopsys)
Physical Design	Physical Compiler & Astro (Synopsys)
Timing Analysis	PrimeTime (Synopsys)
Physical Verification	Calibre (Mentor Graphics)
Formal Verification	Formality (Synopsys)
Dynamic Simulation	VCS (Synopsys)

Figure 5.10. List of EDA tools used for the domino logic Viterbi decoder.

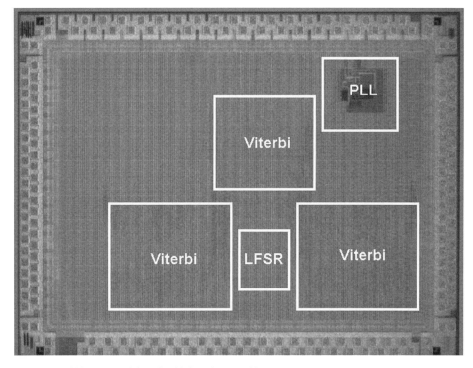

Figure 5.11. Chip photo of the Viterbi decoder test chip.

four-phase clocking scheme, although the design flow supports the use of more or less clock phases. While the actual implementation of the Viterbi test chip used a set of specific tools, the underlying flow is based on Tcl scripts that are largely tool-independent. This ensures that the synthesizable domino logic flow can quickly leverage advances in EDA tools, as is common in static ASIC design. Figure 5.10 lists the different EDA tools used in the implementation of the test chip.

5.4.1 Silicon results

Figure 5.11 shows a photograph of the chip die. The results capture module is not shown explicitly in Figure 5.11 as the hierarchy of the module is flattened to improve its performance. The area of the domino Viterbi module is 0.97 mm by 0.97 mm, leading to a total area of 0.94 mm^2. The Viterbi module is composed of 28,500 standard cells, of which 35% are static. The domino Viterbi module has an area that is 1.4 times greater than the static design. This was despite almost a 100% duplication of logic in the domino module. Our efforts in reducing the area overhead of the domino design had managed to reduce the overhead of synthesizable domino logic considerably!

Silicon measurements showed that at 1.2 V the static logic Viterbi decoder operated at 950 MHz while the domino design operated at 1200 MHz. This is a speedup of slightly more than 1.25×. At 1.56 V, the maximum supply voltage, the static design operated at 1.2 GHz while the domino design operated at 1.55 GHz, giving a speedup of about 1.3×. Timing comparisons using CAD tools had suggested that the domino design was 1.5 times faster than the static design. There are a number of explanations for this discrepancy, the most probable reason being that limited tester availability meant that the maximum operating frequency for the domino design was finally tested on only three randomly picked die on the test wafer. The maximum operating frequency for the static module was found by testing all functional die on the wafer. The integrity of the on-chip power distribution network may also have impacted the maximum operating frequency of the chip. Finally, the maximum operating frequency for the four-phase PLL was limited to about 1.6 GHz. This may also have limited the operating frequency of the domino design at the higher supply voltage level.

The power dissipation in the static and domino designs was also measured. At 800 MHz the static design dissipated 46.3 mA while the domino design dissipated 75.3 mA. The power dissipation in the domino design is 1.63 times greater at the same frequency. The power dissipation was reduced by the use of static cells in the domino implementation.

The functionality and speedup of the fully synthesizable domino logic implementation indicate that the approach used by us was reasonable. While the performance did not achieve the full speedup anticipated, there do not appear to be any fundamental roadblocks to achieving that. The area and power overhead of domino logic was also quite well contained and close to our goal of a 1.5× area and power penalty.

5.5 Intel's published domino logic synthesis flow

The published literature on domino synthesis is concentrated on CAD issues related to synthesis of domino circuits, emphasizing primarily the need to construct unate logic structures. Only one published paper describes a full industrial domino synthesis scheme. This paper, perhaps not surprisingly by a group of Intel engineers [3], was presented at the 2002 IEEE International Conference on Computer Design. In this section that paper is discussed, with the approach used by the authors compared with that described in the book till now.

The paper, by Chappel *et al.*, begins with a broad overview of their goals. They state that while domino logic may only be used in a small portion of a chip, it places a high burden on circuit design resources and overall risk. This is similar to our own motivation for pursuing domino logic synthesis. For the Intel designers the purpose of domino logic synthesis is primarily to produce designs that have a performance comparable with custom design. The primary benchmark is thus whether domino synthesis can achieve results similar to those of an experienced human designer. Our own approach to domino has been from the perspective of seeing how to use domino in the context of a traditional static ASIC design methodology. The published approach thus emphasizes a domino synthesis flow that supports the highest performance custom design needs. Indeed, the flow was used to design 2 GHz production modules in a 0.18 μm CMOS process. Since the paper discusses a full domino synthesis flow, from an RTL specification to a final layout, the authors note that their own experience differs from the published literature in their need to develop a domino logic standard cell library and also in coping with the timing complexities of domino logic. These two issues also loomed large for us, as Chapters 3 and 4 of this book testify.

The described approach to domino synthesis follows a standard ASIC design flow: logic synthesis, timing optimization, placement, and routing. The flow was, however, implemented with custom EDA tools and flows. Using custom synthesis tools allowed the designers to achieve their desired circuit implementation. For example, in the initial synthesis a larger set of AND and OR primitives was used than typically found in synthesis modules. This ensured that very wide OR logic structures, which can be efficiently implemented with domino logic, would be present in the design. While standard EDA tools may have been used to implement some tasks in the flow, it is not obvious on reading the paper where this would have been. The unate steps and the timing optimization are not described in great detail either, although the authors do state that they did unate the design to avoid the penalty of using dual-rail domino cells. The final placement and routing appear to be much closer to a structured custom approach than a pure automated ASIC flow. The paper describes the use of only the bottom three metal levels for all block-level routing. The physical design process uses standard cells in which the acceptable noise level at each input has been characterized. This information is used during the timing optimization step and in physical design. The cells are also characterized with timing constraints for the precharge and evaluate phases. Since the timing characterization flow is not described, it is not known. The domino logic standard cell library is described as a "large library of high fan-in complex domino gates".

The results described in the paper are impressive: 2 GHz modules in a 0.18 μm process and synthesis results that frequently exceeded the results obtained by seasoned designers [3]. Since the comparison point is with custom domino implementations, it is not possible to know how the results compare with static synthesis. The synthesis approach appears to be tied closely to a very detailed partitioning of the microarchitecture. For example, the synthesis examples mentioned in the paper use less than 1000 cells. The relatively small module sizes, taken together with the need to use only three routing layers, suggests that domino synthesis was used primarily to construct small blocks which are then assembled to produce larger blocks. This is possible since presumably the design has been carefully partitioned in terms of functionality and timing constraints. The authors do not describe the product in which this domino-synthesized module is used, although the clock frequency and year of publication of the paper suggest that it was in some version of a Pentium 4 processor.

5.6 Conclusions

The design examples in this chapter show that domino logic can effectively be incorporated in ASIC design flows. The dynamic execution unit showed that it is possible to achieve an excellent speedup in datapath and arithmetic operators by using hand-instantiated domino logic cells. Standard ASIC timing analysis and simulation tools can then be used to verify the design. The silicon results showed that the speedup possible with domino logic was stable across different die and across variations in supply voltage and operating temperature.

The DSP and Viterbi test chips demonstrated that it is possible to use a domino logic synthesis flow to synthesize relatively complex logic modules. While the speedup achieved was lower than what we had hoped to achieve in the Viterbi module, the results are still very encouraging. In light of the fact that timing analysis indicated that better results should have been achieved, it seems reasonable that greater speedup will be possible with a fully synthesizable domino logic flow.

The last paper mentioned here gives the results presented by a group of designers from Intel. While the flow used may be too onerous for most ASIC projects, the paper does illustrate the validity of domino synthesis and also, as with static synthesis, the fact that domino synthesis can often achieve results comparable with custom domino design at much lower cost and design time.

The four case studies presented in this chapter are broadly optimistic about the use of domino logic in an ASIC-style methodology. Still, some questions do remain. Will domino logic continue to be effective as process geometries scale? Are other dynamic logic structures suitable for incorporation in ASIC design flows? What other flow optimizations are possible with domino logic? I will attempt to address these questions in the next and final chapter of this book.

References

1. M. R. Santoro and M. A. Horowitz, SPIM: a pipelined 64 × 64-bit iterative multiplier, *IEEE Journal of Solid-State Circuits* **24**(2), April 1989.
2. D. Harris, *Skew-Tolerant Circuit Design*, Morgan Kaufmann Publishers, San Francisco, CA, 2001.
3. B. Chappell *et al.*, A system-level solution to domino synthesis with 2 GHz application, 2002 IEEE International Conference on Computer Design, Freiburg, Germany, September 2002.

6   Evolution of domino logic synthesis

6.1   The state of digital ASIC design methodologies

Digital ASIC design methodologies are now mature technologies. While EDA tools continue to progress and improve, the basic algorithms on which they are based have been well optimized. In addition, the high-speed needs in an ASIC often tend to be focused on small or medium-sized blocks of logic, while the current focus for EDA tools is on dealing with the massive complexity of systems on-chip. Static logic libraries, like EDA tools, have also improved in the last few years, especially with the introduction of pulse-based flip-flops [1, 2]. Beyond that there does not appear to be very much one can do to improve performance significantly beyond the incremental work of increasing the number of cells and type of libraries provided for the synthesis tool. This is common for many maturing industries, where once the low-hanging fruit has been picked further improvements require considerable effort, often for limited gain.

Before the reader decides to accept the limitations in ASIC design flows with the calm serenity with which it is best to accept the unalterable frailties of the human condition, and other such phenomena, it is perhaps useful to remember that custom designs still remain significantly faster than ASIC implementations in the same process generation [3]. This suggests that there still remains scope for further speed improvements in ASIC flows by using custom design techniques. Once the architectural and logic techniques used in high-speed design are applied, the primary performance limiter in an ASIC flow is its inability to use advanced circuit families. In this book we have described how it is possible to use domino logic in an automated design framework. Some readers may argue that there is no need for a synthesizable domino logic flow, since static logic can provide sufficient speed in current CMOS processes. While this may very well be true for some designs, I would argue that it is often difficult to understand how a different technology would be used if it were available. Let me comment on some applications for a high-speed synthesizable digital design technology:

- As VLSI systems increasingly become systems on a chip, many traditional system-level optimizations are being made on-chip. For example, if using a faster synthesizable domino logic processor core can replace several slower cores with a single one, then the technology becomes appealing. Competitive and pricing pressures reward differentiated, high-value design solutions. The use of domino logic in microprocessors

and custom circuits is possible due to the relatively long design cycles for microprocessors, which allow for time-consuming custom design techniques to be used. If domino logic can be applied using ASIC-style design techniques, it may become much more attractive to other semiconductor market segments.
- If a legacy design or IP needs to be made to run faster, it can be done by implementing it with domino logic. This avoids the costs incurred in redesigning the microarchitecture and porting the software.
- Much of the current research in analog design focuses on the use of very high-speed digital logic to complement and control the analog circuitry. This represents a small, but critically important, part of the total circuitry in the design where domino logic may be useful.
- As leakage power starts to dominate the total power dissipation in scaled CMOS processes, designs increasingly use low-leakage process options. These options invariably result in slower transistors. Domino logic could allow designers to regain some of the performance lost due to using low-leakage transistors. For cost and power budget purposes it is often very appealing to limit designs to using only a single low-leakage process option.
- Arguably, the most powerful way to reduce power dissipation in digital applications is to design the chip to run as fast as possible and then to reduce the supply voltage to achieve lower power at the required frequency [4]. This follows from the quadratic relationship between the dynamic power dissipation in the design and its supply voltage. Thus, reducing the supply voltage by half lowers the power dissipation to a quarter of its original value. Despite its higher intrinsic power dissipation, domino logic could allow for critical parts of a design to operate with a lower supply voltage, and hence power consumption. From a system cost perspective, it is always preferable to use a single power supply for a chip. Domino logic could help ensure that all modules in the chip can use a lower supply voltage.

In this, the final chapter of the book, a number of issues related to the future applicability and evolution of a synthesizable domino logic flow are discussed. Since the performance of circuit logic families is strongly coupled to the underlying semiconductor technology, I start this chapter by trying to see if domino logic will continue to be beneficial as CMOS scales further into nanometer dimensions. Since domino logic needs a number of different clock phases, the chapter also has a short section describing practical approaches to generating them. Subsequent sections of the chapter will look at possible design flow improvements for synthesizable domino logic. The chapter ends with a summary of the reasons that I believe a synthesizable domino logic flow has merit.

6.2 Process trends and domino logic

One of the questions that arises in the discussion about domino logic is its future scalability. A domino cell turns on once the input voltage rises above an NMOS transistor threshold. A static gate starts switching when the input voltage of the cell reaches approximately the Vdd/2 level. As the supply voltage of CMOS processes has scaled from 5 V in 1 μm CMOS processes to 1 V in 65 nm devices, the difference between Vdd/2 and an NMOS transistor threshold voltage has shrunk. This has diminished the switching speed advantage of domino logic. For 90 nm processes and below, the scaling of Vdd does, however, slow considerably due to noise margins. Our own experience with designing domino logic cells at 180 nm, 130 nm, and 90 nm suggested that greater effort had to be expended to achieve a 1.5× speedup in designing domino standard cells as the processes shrank.

Despite these challenges the logic effort advantage of domino over static logic, by which we measure the input capacitance that must be switched for a fixed output drive strength, remains true [5]. This stems from the fact that domino cell inputs have to drive the NMOS transistors only, whereas static cells must drive both NMOS and PMOS transistors. For a fixed input transistor size, a domino cell will have greater drive strength than an equivalent static cell. Further supporting the continued use of domino logic are published reports detailing the continued use of domino logic in microprocessor circuits, indicating that in custom designs there still exists an advantage in using dynamic logic styles [6].

One of the largest changes noticed while working with a number of different CMOS processes over the last few years has been the dramatic increase in leakage currents as processes have scaled. This has been particularly pronounced in high-speed CMOS processes, which have low threshold voltages and hence high source to drain leakage. For such processes, especially under a high operating temperature, leakage current can become the primary power dissipation mechanism. In order to limit the impact of leakage power in scaled CMOS, higher threshold voltages and thicker gate oxides are increasingly being offered. These changes are effective in reducing the leakage through the transistors, although at the price of reduced current drive, and hence lower operating speed. To compensate for the reduction in speed, the nominal supply voltage of low-leakage CMOS processes is often higher than that of a higher speed process. The increased dynamic power dissipation in a "low-power" process is evidently acceptable due to its drastic reduction in leakage current. One of the possible applications of domino logic is to gain back some speed lost due to the transition to a low-leakage process. This becomes more interesting if it can be achieved while using a lower supply voltage.

In the last few years CMOS processes have started using strain engineering to improve carrier mobilities [7]. Moving ahead, the greatest challenges to CMOS appear to be the need to deal with leakage currents and the increased variability associated with atomic-level effects [7]. The problem of curtailing leakage appears to be more tractable, with the projected use of high-k dielectric and metal gates at the 45 nm or 32 nm CMOS node. Process variability seems best resolved with a tighter link between design, especially physical design, and manufacturing [7]. Future evolutionary, or indeed revolutionary process changes, may bring into disfavor domino or other clocked logic styles. This will be particularly true if the relative speed of NMOS to PMOS transistors changes greatly, since domino logic assumes that NMOS transistors are faster than PMOS ones. While technology changes can be abrupt, the economics of a business tend to change more gradually. The sheer investment in CMOS technology up to now means that the current manufacturing capacity will continue to play a crucial role in the semiconductor industry for many years to come. Thus, even if domino logic should lose favor, the sheer volume and diversity of existing CMOS processes, for digital, analog, RF, and high power applications, many of which can be produced very cheaply using fully amortized manufacturing lines, mean that a large number of applications will continue to exist in which domino and other innovative design solutions can be applied. This is particularly true since the fixed cost to make reticle masks in a cutting-edge CMOS process have been increasing very rapidly ($9 million for a 45 nm process has been suggested [8]), making it uneconomic to port applications other than those at the very highest of volumes to the latest processes.

6.3 Clocking methodology for domino circuits

Until now we have assumed the use of skew-tolerant clocking methodology, generally with four clock phases [1]. This scheme has a number of advantages, including relaxing the constraints on the clock tree for the design. This is important since it is very difficult to design large distributed clock trees with automatic place and route tools without incurring a delay penalty. The fault is not necessarily in the clock tree synthesis tools themselves, but as much a consequence of standard design flows. Clock tree synthesis is generally deferred until the design has already been placed. Under such constraining circumstances it is unrealistic to assume that one or more clocks can be distributed to a very large number of endpoints with no skew between the clock arrival times. Even when very tight clock skews can be met it often requires a large number of big clock buffers, which consume a great deal of power.

There are a number of ways in which the four required clock phases can be generated. David Harris has suggested the use of a circuit with a clock source driving three inverters and two inverters to generate a 180 degree phase shift (this will generate phase ϕ1 and ϕ3). The other two clock phases can be generated by delaying these two source clocks. Clock choppers can be used to vary the width of the high clock to achieve greater overlap and skew tolerance [1]. While phase error for the two delayed phases does change with process, voltage, and temperature changes, that should be acceptable. What is needed is to ensure that the slow corner timing is met. While the timing requirements on a particular clock phase may be impacted by process and environmental conditions, this should be more than offset by the speed gained due to the improvements in the transistors. Of course, careful simulations need to be run across different process corners to ensure that the solution remains valid.

An alternative approach to generating four clock phases is to do so directly in the phase locked loop (PLL). In most digital systems the PLL provides a source clock which is used to clock the digital elements. It is possible to generate four accurate clock phases by dividing down the output of a PLL by four. If such a high-speed PLL is not available or suitable, it may be possible to modify the PLL to help it generate the needed clock phases. The heart of the PLL is an oscillator that is generally controlled by an input voltage. The oscillator is consequently called a voltage-controlled oscillator (VCO). In

CMOS circuits there are two common ways to build VCOs: as back-to-back inverters, or using an inductor capacitor (LC) tank. Since an odd number of inverters in a feedback loop will constantly switch phase, this configuration can be used to generate a clock. The capacitive load on the inverters can be altered via a voltage to ensure that the generated frequency is constant across process and environmental changes. In such a VCO structure it is possible to approximate the four clock phases by pulling four different outputs from the VCO. Since the total number of inverters is odd, it is not possible to generate all clocks precisely. An approximate copy of the four clock phases can, however, be obtained. The degree to which these four phases deviate from the ideal depends on the number of back-to-back inverters in the VCO loop. As the number of inverters increases, the error reduces. Unfortunately, when very high speed is needed, the total number of inverters used in the VCO is reduced. This increases the error. Since the skew-tolerant clocking algorithm tolerates skew between the clocks, it should accommodate the slight skew differences between different phases. If the skew between the phases is deterministically known before phase assignment, it is possible to modify the phase-assignment algorithm to support the modified clock phases.

An alternate mechanism to generate a VCO is to use an LC tank. An LC tank forms a natural oscillator, with inherently better jitter performance than a circuit with back-to-back inverters. Since these VCOs typically produce differential outputs, it is possible to divide the output of the PLL by two to generate the four clock phases. Since inductors tend to be quite large, this form of VCO is usually limited to radio frequency (RF) applications. If available for driving a digital circuit, it provides an excellent source for the four clock phases.

When multiple clock phases are used, it is assumed that there are at least a few cells on each clock phase. Having a clocking system with only a single cell on each clock is not very practical. In some designs the maximum operating speed may limit the logic on each pipeline stage to four cells or less. Examples of such blocks include clock divider or data converters. These designs tend to be very fast, perhaps running at a multi gigahertz range in a current CMOS process. As the speed of digital CMOS circuits increases, their potential applications expand to include analog functions, which traditionally have been designed using analog design techniques. The use of digital-centric architectures is driven not only by the availability of faster digital logic, but also by the difficulty in using traditional analog design techniques with the low supply voltage that must be used in a current CMOS process. Pushing the limits of speed for a particular CMOS process (this was once described memorably to me as realizing the "unnatural speed of silicon") requires great effort, often using custom design. If domino logic is to be used for such circuits, it may be impractical to use multiple clock phases. Using the two phases of a single clock source may be more practicable. Extensive simulations need to ensure that the design is stable and without any hold failures. While a formal cell characterization methodology may not be necessary, the same failure mechanisms remain possible and need to be avoided.

One of the difficulties in using domino logic for such deeply pipelined circuits is that the precharge phase starts to become as critical as the evaluate phase. It must be remembered that domino works best when the logic pipelining is such that precharge should not

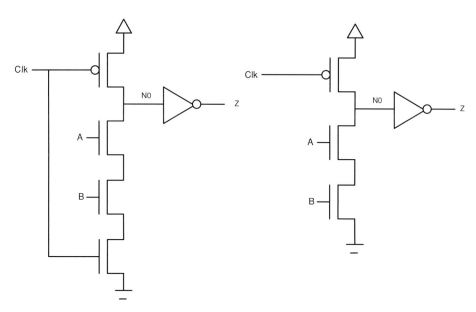

Figure 6.1. A footed and non-footed domino AND gate.

become critical, since the domino logic is most effective when the evaluate phase can be emphasized at the cost of the precharge timing, with no negative consequences. Despite these difficulties, the authors' experience with designing extremely high-speed modules suggests that domino, or some form of dynamic logic, still tends to hold up well against static logic.

6.4 Synthesizing other dynamic logic families

In addition to domino logic there are a number of other dynamic logic styles available in CMOS. As we developed the synthesizable domino logic flow, we investigated the use of some of them. We will next describe some of these logic families and their suitability to be included in an automated design flow.

6.4.1 Non-footed domino logic

In addition to being an upper-class English abbreviation for the game of soccer, footer is also used to refer to the lowest NMOS transistor in a domino pull-down stack. Customarily this transistor is connected to the clock. In Figure 6.1 a non-footed domino AND cell (on the right) is shown along with a footed cell (on the left). It can be seen that the height of the NMOS stack in the AND gate is reduced from three transistors to two with a non-footed design. Since this reduces the number of transistors in series, the delay through the stack is reduced. Typically we have seen delay reductions of at least 10% for a non-footed domino cell compared with the footed version. Another advantage of non-footed domino is that the clock load is reduced, since it does not need to toggle the

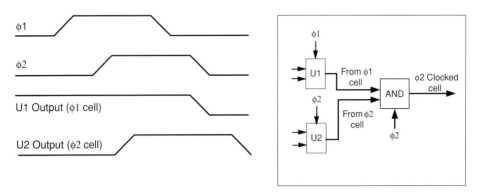

Figure 6.2. Illustrating the condition where a footed domino cell can be substituted for a non-footed one.

footed transistors. Since the clock is the largest source of power dissipation in domino designs, this reduction in power dissipation is most welcome.

The primary difficulty in using non-footed domino cells is that a race exists between the clock and the inputs in such cells. Specifically, when the clock turns on for precharge, the NMOS pull-down network must be turned off. If this does not occur, a shortcircuit current will lead to unnecessary power consumption. Since the power dissipation is directly between the Vdd and ground rail, it will be large. Furthermore, if the NMOS pull-down network is not shut off once the cell enters precharge, the time at which the internal node actually starts to precharge is delayed. If the node is not properly precharged by the time the evaluate cycle starts, a functional failure can happen. Since the inputs to the NMOS chain usually come from other domino cells, they are delayed by the precharge delay of the previous cell. This makes the precharge time for a non-footed domino cell, following a footed one, often critical. Sizing the transistors in the driving cell to improve precharge delay can reduce precharge delay, although this leads to greater evaluate delay.

The need to achieve acceptable precharge and evaluate delays makes it difficult to merge footed and non-footed domino cells for very deeply pipelined designs. For designs with many logic delays between registers, it is easier to use footed domino cells since the extra precharge delay is more likely to be acceptable. Whatever the logical structure of the design, there does, however, exist one case where it is always possible to substitute a non-footed domino cell for a footed one. If at least one input for each NMOS pull-down stack in a domino cell comes from a domino cell in an earlier phase, while another input comes from a cell in the same clock phase, then substituting the cell with a non-footed equivalent can be done safely. This is shown in Figure 6.2, where cell U1 is clocked by an earlier phase and U2 is clocked by the same phase as the cell under consideration. The need to have one input clocked by the cell in the same phase is to avoid the condition where the inputs to the cell are high when the cell precharges.

To cause minimal timing modification to the cell in a design it is safest to substitute footed cells with the non-footed equivalents only after the cells have been placed. Such a post-placement substitution will reduce the power dissipation in the design on top of

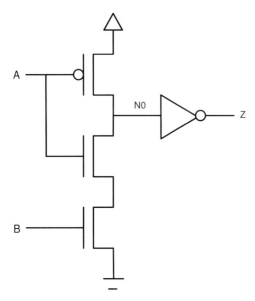

Figure 6.3. A data-driven dynamic logic (D^3L) two-input AND cell.

the speed improvement. Unfortunately, it does not reduce the area of the design since the placement of the other cells is frozen. Our own simulation experiments on a test circuit unfortunately found only a limited number of cells in which a footed domino cell could safely be replaced by a non-footed candidate. The power and delay savings did not warrant the complexity in developing complete non-footed domino libraries.

6.4.2 Data-driven domino

Figure 6.3 shows a data-driven domino logic AND2 input cell [9]. This logic family is in fact an incomplete static logic cell since no clock input is provided to it. Data-driven domino cells have evaluate delays as fast as a non-footed domino cell, with lower power dissipation, since there is no clock transistor. The disadvantage of this logic family is the very long precharge delay. The logic is also difficult to implement for non-AND-type cells. This is because those cells need series PMOS transistors to control the precharge path, further slowing down precharge. The extra precharge delay and the limited number of possible standard cells which can effectively be implemented with this logic style made us not use it in a synthesized design flow.

6.4.3 Compound domino

Another dynamic logic style that we experimented with during the development of our domino logic synthesis flow was compound or complex domino [10]. Figure 6.4 shows a compound domino logic cell implementing a four-input AND function. The output inverter in the cell is replaced by a static NOR cell. This means that five NMOS transistors do not need to be stacked for the pull-down chain. Instead, two stacks of

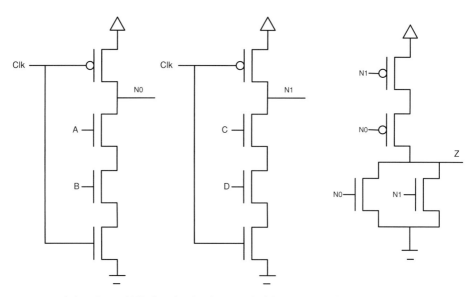

Figure 6.4. A four-input AND function implemented with compound domino logic.

three NMOS transistors in series are connected via a NOR cell. Compound domino can be used to implement very complex domino logic cells. Unlike the other logic styles discussed till now compound domino, when used with NOR-type output drivers, does not add a significant precharge delay to the cell. If the output driver cell is a NAND gate then the compound domino cell does lead to longer precharge delays, due to the output signal needing to traverse through an extra NMOS transistor. Unfortunately, our synthesis experiments showed some limitations in using compound domino structures. Firstly, compound domino logic is best suited for implementing large, relatively complex standard cell functions. The synthesis tools were not, however, using these cells as much as expected. This may have reflected a tool limitation, but it did still negate the extra effort needed to design them. Secondly, we saw that to ensure a fast evaluate delay in the design, very large output PMOS transistors were needed in the NOR output driver. This meant that the compound domino cells ended up becoming much larger than regular domino cells.

6.4.4 Other dynamic logic styles

There are a number of other varieties of domino logic that are discussed in the literature. While we did not try to use these logic styles in the synthesizable domino logic flow, I will still briefly discuss their suitability towards being used in a synthesizable domino logic flow.

Zipper CMOS and NORA logic are two related dynamic logic styles in which the output inverter is removed. On the surface this seems reasonable, since the inversion consumes delay and energy without performing any other logic. The presence of an extra inverter does to a degree hamper domino logic. Indeed, one has to be careful

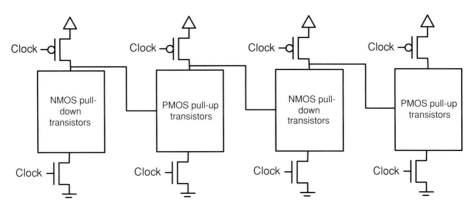

Figure 6.5. Zipper logic uses alternating groups of pull-up and pull-down transistors.

when comparing static and domino logic designs to ensure that full critical paths are studied and that one does not merely compare a domino design with an identical static one. Static logic tends to use inverting logical functions, especially NAND gates, quite liberally, of which there is no domino equivalent. Synthesis experiments in which we did not allow the use of inverting logic other than an inverter with a static logic library showed a penalty of approximately 20%. In defense of standard domino logic, it should be stated that while extra inverter delays are encountered, all computational functions involve traversing only through NMOS transistors. For Zipper logic, critical paths will need to go through PMOS transistors.

In order to ensure that Zipper logic can operate correctly, NMOS pull-down segments are followed by PMOS pull-up segments. This is shown in Figure 6.5. This ensures that the output of the NMOS logic does not inadvertently discharge the next logic stage. The removal of the inverter, however, makes the logic far more susceptible to noise, since the inverter helps filter out noise at the input of the cell. Susceptibility to noise makes Zipper logic unsuitable for use in an automated design framework, since it will be difficult to control the noise on individual nets. Zipper logic has not been used widely in industry [10], which would appear to be to do with its limited noise tolerance.

Self-resetting domino was the circuit logic style used to design the first microprocessor test chip that operated at 1 GHz [10]. In order to control clock skew, the clock travels with the data. A replica circuit delays the clock at each stage as much as the data is delayed in the cell. Since clocks tend to be generated locally and not distributed globally, a local feedback mechanism is used to reset every domino cell after the evaluate phase. Figure 6.6 shows a self-resetting domino cell. This logic style is very elegant due to its use of some asynchronous concepts in a clocked logic style. Locally distributing clocks is equivalent to having many clock phases. For custom datapath structures built as tree-type structures it seems reasonable to be able to match the delays across different cells. The challenge with self-resetting logic occurs when the inputs to a cell have different arrival times, something that is almost guaranteed to occur with an ASIC implementation. Delay modules will then need to be set variably for the different inputs to each cell. Such a solution does not appear tractable at present with an automated design solution.

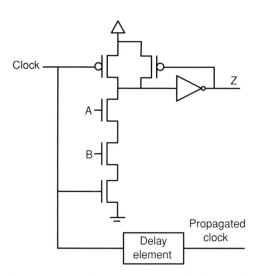

Figure 6.6. A two-input self-resetting domino AND cell.

In 2004 a group of researchers at Intel described a new class of logic called low-voltage swing domino [11]. The logic uses differential inputs and a pass gate NMOS structure. Small voltage differences are sensed in a sense amplifier. The advantage of the logic is that very complex logic structures can be implemented, with very small voltage swings causing the output of a cell to switch. This reduces the power dissipation in the design. While this is an interesting approach, the technique seems unsuited for automation due to the use of very complex cells which are difficult to map to and the need for an inordinately large standard cell library.

Despite some advantages that other dynamic logic styles enjoy, standard domino logic continues to be the easiest technology to use in a synthesizable design flow. After standard footed domino, non-footed domino may be the easiest to use due to the ease with which a non-footed cell can be replaced by a footed one. If done correctly, each substitution leads to a faster evaluate delay and less power dissipation, both good things. The reader is reminded that the primary concern we had with many of the dynamic design styles mentioned relates to their suitability to be applied in a synthesizable solution. If schematic capture is used, more aggressive design styles may be used safely and advantageously.

6.5 Flow improvements for domino synthesis

When comparing a synthesized domino implementation with a static one, the extra area penalty, with its attendant power dissipation, is often the most unacceptable compromise required to allow the domino logic to be synthesized. Consequently, there has been a considerable degree of interest in trying to create domino logic-compatible designs in which duplication can be avoided. An intern who worked with us one summer became quite committed to finding a solution to the problem of avoiding the need to duplicate logic

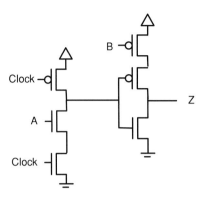

Figure 6.7. A binate dynamic logic cell.

in domino implementations. Throughout the summer he produced a set of ideas on how to potentially avoid duplication. Unfortunately, none of the proposed solutions worked. Indeed, some published papers have proposed similar solutions to avoid duplication. In order to provide an example of the kind of structures proposed to allow inverting logic in domino cells, we refer the reader to Figure 6.7.

Figure 6.7 shows a domino cell with input A driving the pull-down NMOS stack and input B being connected directly to the output PMOS drive stage. When the clock is low the output of the cell is 0, ensuring that the cell precharges correctly. Furthermore, if the inputs A and B are 1 and 0, respectively, the output turns high. This suggests that the cell correctly implements the function AB'. Indeed, the cell does implement the function correctly provided that input B is zero or that input A always changes after B has risen. However, if input B changes after input A, the output of the cell incorrectly settles to 1. Since this is a dynamic logic style the subsequent rise in input B, which should cause the output to drop to logic 0, cannot reverse the output value, leading to a functional error.

If input B always arrives before input A, it is possible to use a circuit such as this to remove the inverter without duplicating the full cone of logic for input B. This case on its own is somewhat trivial since the inverter can easily be removed for this gate by bubble pushing. However, this gate could be part of a logical structure where a trapped inverter is present. Under those circumstances the ability to use timing relationships to remove the need for full duplication starts to become very interesting. A generalized scheme for doing this is discussed in a paper described next.

6.5.1 Allowing binate logic in domino designs

At the Design Automation Conference held in 2004, Cao and Koh proposed a solution for avoiding duplication wherever it is possible to gate a fast-arriving inverting input by a slow-arriving non-inverting one [12]. In Figure 6.8, input A arrives before input B in a domino logic AND gate. The inversion here is performed with a static inverter, although it is also possible to use a circuit such as that shown in Figure 6.7 to perform the inversion. Cao and Koh state that if precise delays are known, the redundant unate logic for an input can be reduced to the binate form. Stripping the redundant logic reduces

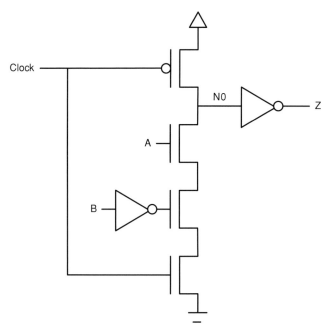

Figure 6.8. Known timing relationships between cell inputs allow inverting logic.

the power dissipation in the design but does not reduce the area, since it has already been placed. To reduce area the authors propose a compaction scheme to reduce the area without significantly altering the timing relationships in the design. The authors provide encouraging data from their approach. The average reduction in power for their scheme is 30%. With a compaction scheme the area is reduced by 20%, although the power increases. While these results are positive, a few words of concern need to be stated. The authors do not consider a detailed clocking scheme for the design. Our own experience suggests that in order to have correctly functioning logic, the input arrival times at domino cells need to be made close to each other (by inserting domino or static buffers if needs be). This process will greatly limit the number of candidates in which the maximum arrival time of an input is less than the minimum arrival time of another input, especially after clock skew is considered. Furthermore, as the logic depth of a design increases, there tends to be a spreading of the difference between the minimum and maximum arrival time of an input signal to a gate. This follows from the increase in the number of paths to the gate inputs. For complex designs this means that the proposed scheme may be most applicable for cells close to the primary inputs, but less so further downstream. For the design considered by the authors (C432 through C7552 from the ISCAS benchmarks), the maximum number of cells in the design is 3095. These are relatively small modules. Finally, the authors limit themselves to domino AND gates, OR gates, and buffers. The approach needs to be tested against a much larger library of domino cells. For small blocks, especially in custom or structured custom environments, the proposed idea seems useful.

6.5.2 Clock modification to allow for the use of more non-footed domino

Another possible flow modification that has promise in a domino synthesis flow is to modify the width of some clock phases in order to ensure greater use of non-footed domino logic cells. In order to understand the purpose of this scheme, let us revisit the issue of what the clock phases for the inputs of a footed domino cell must be, so that the cell can safely be substituted for a non-footed one. For our purposes we will consider a two-input domino AND gate. If the cell is clocked on phase $\phi 2$ and both of the inputs come from domino cells clocked on phase $\phi 1$, then a period of time exists when clock phase $\phi 2$ is zero and clock phase $\phi 1$ is high. During this period the cell is in precharge, but may not be able to complete precharge if the inputs of the cell turn on the pull-down NMOS stack. A failure to precharge will mean that the design is functionally incorrect. Even if this does not happen, the possibility for severe shortcircuit power dissipation exists. If, alternatively, the two inputs are also clocked on phase $\phi 2$, then we could also face problems since the inputs to the cell only turn low after the driving cell has discharged. This slows down the time at which precharge starts, again possibly leading to a functional error. The possibility of large circuit currents also exists after the phase clock turns low, since some inputs will remain active for a while. For this reason non-footed domino can only safely be applied when one input is clocked by an input on phase $\phi 1$, which ensures that the cell will start to precharge immediately after phase $\phi 2$ turns low and another input is clocked by phase $\phi 2$, which ensures that the cell does not precharge when phase $\phi 1$ turns high but phase $\phi 2$ does not. The rule can be generalized as requiring that each input to every pull-down network has at least one serial NMOS transistor clocked by a cell on the same phase and another transistor that is clocked on an earlier phase. We found that, for our test cases, very few cells met these stringent conditions.

In order to allow for more non-footed cells to be replaced by footed ones, a modification in the clock phase widths has been suggested [13]. The idea is as follows: each phase of the clock is generated with two different high pulse widths, a regular 50% duty cycle phase, and another phase whose high pulse width is reduced by an amount of time equal to a typical precharge delay and clock skew. This number depends, of course, on the process technology and library design. The narrower high pulse phase is then used to clock the first cells on the clock phase. The advantage of using such a narrower clock phase is that the output of these cells can then be used to clock other domino cells on the same clock phase, with it being possible to safely switch the second cell for a non-footed version. This is shown in Figure 6.9, where a cell on phase $\phi 1$ with a narrower high pulse is used to clock another domino cell on phase $\phi 1$. Since the narrower pulse high causes the output of the cell to enter precharge earlier, the output from the cell should be low when a cell it drives is on the same phase but with a wider high pulse that turns low. This ensures that the second cell, which only has input coming from cells on the same phase, can still be switched with a non-footed version. Of course, detailed timing analysis needs to be performed to ensure that the substitution is acceptable for every particular cell where it is to be applied. The technique can be extended to consider a number of high pulse widths for clocks. Every time a clock with a narrower

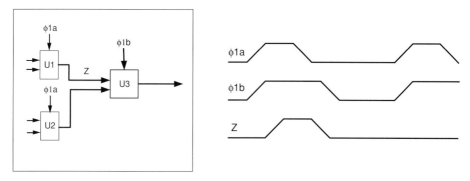

Figure 6.9. Using narrower clock pulses allows for more non-footed domino cells.

pulse feeds a domino cell with a wider phase clock, it is possible to use a non-footed domino cell.

6.6 The case for domino logic synthesis

The demand for high-speed ASIC design is driven by the needs of ASIC designers seeking higher performance and custom designers seeking higher productivity [3]. The increasing costs of designing and manufacturing an ASIC have meant that there has been a dramatic reduction in the number of ASIC design starts over the last few years. From 7700 design starts in 2000, the number is expected to drop to 3200 in 2007 [14]. On top of the $9 million masks costs for a cutting edge 45 nm design, there are the design costs of between $20 and $50 million [8]. These massive costs mean that chips must transition from being point solutions to becoming platform-based solutions that can address broader markets [3]. For solutions such as this, there is a great demand for high performance to justify the costs and time needed to produce new products. A synthesizable domino logic flow can provide such a differentiating performance advantage.

Once available a synthesizable domino logic technology provides ASIC designers with access to a cost-effective, if less capable, form of domino logic. This technology could profoundly broaden the use of domino logic. Clayton Christensen has argued that disruptive technology involves taking a sophisticated technology and making "things simple and low-cost so that a whole new population can own things and do things" [15]. It is possible to imagine a scenario where domino synthesis becomes progressively more sophisticated, finally allowing ASIC designs and system architects easy access to the benefits of using custom designers for whatever block needs it. It can also become a tool to help custom designers be more productive, allowing them to avoid basic domino design, and focus instead on more novel design solutions and other critical circuitry.

Of course these predictions assume that an industrial-strength domino logic synthesis flow can be constructed. I believe that with current EDA tools it is possible to deploy a reliable domino synthesis automation framework. Such a technology provides a very promising avenue to improve the performance of digital systems implemented as ASICs.

References

1. D. Harris, *Skew-Tolerant Circuit Design*, Morgan Kaufmann Publishers, San Francisco, CA, 2001.
2. C. Bittlestone, A. Hill, V. Singhal and N. V. Arvind, Architecting ASIC libraries and flows in nanometer era, 40th Design Automation Conference, Anaheim, CA, June 2003.
3. D. Chinnery and K. Keutzer, *Closing the Gap between ASIC and Custom: Tools and Techniques for High Performance ASIC Design*, Kluwer Academic Publishers, Norwell, MA, 2002.
4. D. Markovic et al., Methods for true energy performance optimization, *IEEE Journal of Solid State Circuits* **39**(8), August 2004.
5. I. Sutherland, B. Sproull and D. Harris, *Logical Effort: Designing Fast CMOS Circuits*, Morgan Kaufmann Publishers, San Francisco, CA, 1999.
6. S. P. Wijeratne, A 9 GHz 65-nm Intel Pentium 4 processor integer execution unit, *IEEE Journal of Solid-State Circuits* **42**(1), January 2007.
7. T.-C. Chen, Where CMOS is going: trendy hype vs. real technology, *IEEE SSCS Newsletter*, September 2006.
8. M. LaPedus, Cost cast ICs into Darwinian struggle, *Electronic Engineering EETimes*, Issue 1469, April 2, 2007.
9. R. Rafati, S. M. Fakhraie and K. C. Smith, Low-power data-driven dynamic logic (D^3L), IEEE International Symposium on Circuits and Systems, Geneva, Switzerland, May 2000.
10. K. Bernstein et al., *High Speed CMOS Design Styles*, Kluwer Academic Publishers, Norwell, MA, 1998.
11. D. J. Delagenes et al., LVS technology for Intel Pentium 4 processor on 90 nm technology, *Intel Technology Journal* **8**(1), February 2004.
12. A. Cao and C.-K. Koh, Post-layout optimization of domino circuits, 41st Design Automation Conference, San Diego, CA, June 2004.
13. R. Mader and B. Bourgin, Unfooted domino logic circuit and method, *US Patent Number 7233639*, June 2007.
14. M. LaPedus, IBM aims to revive ASIC with next-gen spin, *Electronic Engineering EETimes*, Issue 1479, June 11, 2007.
15. B. Fuller, Make disruption work for you, prof duo says, *Electronic Engineering EETimes*, Issue 1374, June 6, 2005.

Index

adder
  carry propagate 25
  carry save 25
  floating point 28–29
  Kogge–Stone 23
alpha microprocessor 18
arithmetic module 22
ASIC methodology
  design costs 141
  example 4
  in consumer market 21
  total design starts 141
average selling prices (ASP) 21

bipolar logic 2
bubble pushing, see unate transform

characterization
  data pin setup falling 59
  domino cell
    falling delay 53
    rising delay 52–53
    transition characterization 54
  domino register
    delay measurement 64
    hold 65
    scan output delay 64
    setup 65
  hold falling 58
  input pin capacitance 54
  maximum data pin crosstalk 61–62
  minimum pulse width high and low 59
  minimum pulse width high overlap 55–57
  setup rising with respect to clock falling 54–55
  simultaneous crosstalk and charge sharing 62
charge sharing, checking 62
clock
  creating four phases
  hard timing edge 75
  skew 14
  soft timing edge 15
  two-phase for domino 12–14

CMOS
  45 nm process 130
  history of 1–4
  manufacturing capacity mix 129
  power dissipation 2
  process trends 129
  scaling 19
  static NAND gate 1
coupling capacitance 51
crosstalk
  fixing 105
  maximum voltage spike check 105
custom design
  benefits 4–5
  in microprocessor 21
  optimizing across logic and circuit design 29

design rule check (DRC) 105
domino
  ASIC flow
    allowing binate logic 138–139
    applications 127–128
    benefits 70
    challenges 70
    clock tree synthesis 104
    crosstalk fixing by router 104
    design guidelines 91–92
    disruptive technology 141
    dynamic simulations 105
    formal verification 103
    initial placement 98
    non-footed domino 134, 140–141
    overview 72–73
    physical design 103
    portability across different EDA tools 106
    pulse-based analysis 106
    RTL guidelines 95
    silicon results 126
    standard tool-based 71
    synthesis constraints 92–95
    synthesizing other dynamic logic families 132
    uses of domino design 16
    variables 87–91

domino (*cont.*)
  DSP chip
    results 119
    system advantages in using domino 119
  logic
    advantages 15
    AND gate 5
    avoiding explicit flip-flops with 14
    charge sharing 49–50
    clock 6
    clocking techniques 12–15
    compound domino 134
    crosstalk noise 50–51
    data-driven 134
    disadvantages 15–16
    evaluate phase 6
    evaluate transistors 49
    factors to consider before using 35
    full timing model 66–67
    future scalability 128
    implementing binate functions with 9–11
    improving charge-sharing tolerance 63
    improving precharge delay 63
    is dynamic attribute 66
    keeper transistor 49
    lack of contention 8
    low frequency and voltage operation 117
    maximum operating frequency 131
    maximum precharge delay 49
    non-footed domino speed advantage 132
    precharge check 63
    precharge delay 49, 52–54
    precharge phase 6
    precharge transistors 49
    schematic capture 106–107
    self-resetting 136
    speed advantage 6–8
    uninverting nature 8
    Zipper/NORA 134, 135
  synthesis, *see* domino ASIC flow
dual output domino
  advantages 110
  disadvantages 110
  example circuit 109

EDA history 4

fan-out of four (FO4) 20
flip-flop
  D-to-Q delay 41
  hold time in 14, 65
  master–slave 41
  pulse 41
  setup and hold measurement 44
Frank Wanlass 1

glitching 9

high-performance microarchitecture 22–29
hold time
  definition 14
  of pulse flip-flops 41
hot cell 30

integer execution unit chip
  chip description 115–116
  clock generation 116
  crosstalk failure 113
  crosstalk simulation 112
  data flow 108
  datapath 111
  design flow 108
  domino cells 111–112
  domino inverter P/N sizing 115
  domino keeper sizing 115
  dynamic cell P/N sizing 114
  overview 108
  physical design 116
  precharge transistor sizing 115
  silicon results 116–119
  test methodology 116
  transistor sizing simulation 112
Intel
  4004 2
  8088/8086 3
  domino synthesis paper 124–125
  low voltage swing domino 136

layout versus schematic (LVS) 105
logic
  adder and shifter module 29–30
  predictive comparator following addition 26
  self-loading effect 6
  speculative operation 27
logic synthesis
  description 4
  of complex datapath 25

memory
  6T cell 33
  decoder 34
  interface to domino logic 35
  layout requirements 33–34
  sense amplifiers 34
  SRAM example 31
  timing models 31
  using in ASIC design 31
microarchitecture definition 22
microprocessor
  performance predictions 18
  speed evolution 18

Index

minimum pulse width high overlap
  description 104
  iterative fixing 104
  violation report 104
multiplier 24–25

NMOS
  NAND gate 2
  speed disadvantage 3

pass transistor logic
  in standard cells 40
  XOR 40
phase assignment
  challenges 76
  clock width high check 83
  clock width low check 84
  definition 75
  detailed description 101–103
  domino input ports 82
  fan-in phase differences 80–81
  maximum negative slack 101
  mixing static and domino cell 86
  multi cycle paths 84–85
  phase skip limit 101
  phase skipping 78–79
    with slack 101
    with static cells 100
  precharge failure 83
  requirements 77
  simplification of latched outputs 90
  static input port 81–82
  static output port 83
  unbalanced 79–80
  using mixed registers 100
  using skewed clocks 100
phase locked loop 130–131
pin under test 52
pipelining
  deeply in microprocessors 19–20
  limits of 21

process, using worst-case corner in design 120

radio frequency (RF) 11

setup time 14
standard cell
  domino logic compatibility 66
  domino logic well 66
  drive strengths 46–48
  layout 42–43
  library performance versus size 46
  static cell library 127
  timing assumption 44
  track 45
  typical library size 45
switching point 8

timing model, pseudo-static 72
timing verification 105
transistor sizing
  optimal P/N ratio 38–39
  scaling 37–38

unate transform
  binate functions 75
  entire domino library 74
  incremental optimization 97
  output phase optimization 96
  overview 73–75
  removing trapped inverters 74
  static port specification 97

Viterbi decode chip
  description 121–122
  design flow 122–123
  silicon results 124
voltage-controlled oscillator
  back-to-back inverter 131
  LC tank 131